THE WEATHER SOURCEBOOK

....

YOUR ONE-STOP RESOURCE FOR EVERYTHING YOU NEED TO FEED YOUR WEATHER HABIT

Second Edition

by
Ronald L. Wagner
and
Bill Adler, Jr.

Old Saybrook, Connecticut

Dedication

From Ron: To Michael, Rich, Lisa, and Jamie

"The World's Oldest 'Child'" on page 90 appeared originally in the *American Weather Observer* under the title *"It's Back!!!"*

Interior photo credits: Pp. v, 9, 11, 32, 33, 36, 38, 40, 42, 45, 48, 50, 51, 73, 98, 100, 101, 112 courtesy of National Oceanic and Atmospheric Administration; pg. 109 courtesy of Chris Hayes Novy; pg. 136 courtesy of Satallite Data Systems, Inc.; pg. 158 courtesy of Steve Baldwin Photography; pg. 202 copyright © Jack "Thunderhead" Corso

Copyright © 1994, 1997 by Adler & Robin Books, Inc.

All rights reserved. No part of this book may be reproduced or transmitted in any form by any means, electronic or mechanical, including photocopying and recording, or by any information storage and retrieval system, except as may be expressly permitted by the 1976 Copyright Act or by the publisher. Requests for permission should be made in writing to The Globe Pequot Press, P.O. Box 833, Old Saybrook, CT 06475.

Library of Congress Cataloging-in-Publication Data

Wagner, Ronald L.
    The weather sourcebook: your one-stop resource for everything you need to feed your weather habit / by Ronald L. Wagner and Bill Adler, Jr. — 2nd ed.
        p.   cm.
    Includes bibliographical references and index.
    ISBN 1-7627-0080-7
    1. Weather. 2. Meteorology.   I. Adler, Bill, Jr. II. Title.
QC981.W24 1997
551.5—dc20   97-4475
                                                                                      CIP

Manufactured in the United States of America
Second Edition/First Printing

Contents

Introduction ix
1. History 1
2. Weather Lore 17
3. Violent Weather 31
4. Atmospheric Pressure 55
5. Wind 67
6. Temperature 81
7. Moisture 97
8. Clouds 109
9. Marine Weather 121
10. Weather for Pilots 135
11. How to Become a Weather Forecaster 147
12. Major Equipment 159
13. How to Get Weather Forecasts and Information 173
14. Books about the Weather 187
15. How to Learn More about the Weather 197
    Weather Terminology 211
    Index 215

Acknowledgments

Thanks to Lisa M. Kauffman, Denise Kauffman, and Leah Nelson for editorial assistance, research, and good advice. We couldn't have done it without you.

Introduction
Why We Are So Fascinated by Weather

More than thirty years ago, at the peak of Cold War tensions, the United States and the Soviet Union were engaged in a frenzy of activity to increase their capability to launch a barrage of nuclear-tipped intercontinental missiles at each other. Yet even while gearing up for mutual assured destruction, both found a basis for agreement on at least one topic: the weather.

In 1963 the two nations set up a cooperative weather-research exchange line. Connected twenty-four hours a day, the line exchanged photographs of cloud formations gathered from satellites. Even back then the volatile politics attached to missile technology could be overcome by a mutual concern over the one thing that was larger than either superpower.

Nothing else brings people together like the weather. On beautiful spring days neighbors meet neighbors. Children play outdoors and make new friends. Lovers walk hand-in-hand in parks. In the aftermath of a devastating hurricane, people risk their lives to help others to whom they never would speak otherwise. One severe storm can unite enemies, and destroy entire cities.

The weather touches and activates every human emotion and, at one time or another, affects nearly everything we do. The brother–sister singing duo The Carpenters linked the weather and the blues in a song that echoes the emotions of millions of people who agree that "Rainy days and Mondays always bring me down." Buddy Holly used the weather to express his sadness over losing a love by singing, "It's raining, raining in my heart." Describing fog as arriving on little cat feet, Carl Sandburg ascribed living qualities to an otherwise dreary and benign weather phenomenon.

Though fog can be depicted romantically by a gifted poet, it's the worst fear of a pilot whose plane is running low on fuel. Obscuring towers, mountains, and trees, fog is just a gray, wet mask hiding the safety of a runway. After a plane has flown hundreds of miles uneventfully through clear skies, a little patchy, low-lying fog can be fatal to pilot and passengers. At best, fog will delay a flight's arrival.

The weather has altered the outcome of wars. The U.S. Civil War began with the Battle of Bull Run in Virginia. The Union Army—with 35,000 men under the command of General Irvin McDowell—anticipated fighting a quick rout of the Confederates—with only 20,000 men under the command of General Pierre G. T. Beauregard—and ending their uprising before it got started. The month of July 1861 had been drier than normal, however, enabling 9,000 Confederate reinforcements under the command of General Joseph E. Johnson to arrive from the Shenandoah Valley faster than expected. The arrival of such numbers of fresh Confederate troops led to an upset rout that sent the Union Army into a panic retreat to Washington. With the Confederates heartened by their victory, it would take four more bloody years for the Union Army to halt the uprising.

It's only natural that we are fascinated by a power that we cannot control and that affects our activities and emotions so strongly. This inherent fascination makes weather the number-one topic all over the world. Since we cannot control the weather, we've had to settle for the next best thing: studying and tracking it so we can predict its impact on our future. If we have an accurate prediction of coming weather, we can take precautions to minimize its negative effects and make plans to maximize our enjoyment of its most pleasant aspects.

Doing Something about the Weather

If you'd like to get as close as you can to "doing something about the weather," this book is for you. Each chapter is dedicated to a different aspect of the weather, showing how it influences our lives and how you can learn more about it.

Throughout the chapters you'll find selections that are like a "best of" recording by your favorite musical group. Weather experts from all over the United States have contributed articles and sidebars that will help you learn more about weather or entertain you with fascinating stories about the impact that weather has on us all. Finally, each chapter ends with a summary of weather products related to its topic.

The product listings and the information in the collection of articles we've included can be your foundation for becoming a weather expert yourself. Even though you still will be able to do little more about the weather than watch it in awe, this book will help you appreciate more than ever the magnificent weather shows that nature presents and help you understand more of the subtleties of the small yet amazing things that the weather offers daily.

Chapter 1
History

Picture the Earth from a distance, as you've seen it in photographs taken from space. Now envision that picture as the size of a large schoolroom globe. The beautiful blue coloring that comes to mind is the result of sunlight reflecting off the atmosphere that surrounds the Earth.

How much larger would you guess the globe would be if it were surrounded by a scale model of our atmosphere? You would barely notice the difference, because if you added a model atmosphere to a model globe, the life-giving layer of gases surrounding our planet would be only as thick as a layer of clear varnish. Yet this incredibly thin coating provides all our weather, all our life support, and all our protection from the dangers of the sun's radiation and from drifting space objects with which our planet collides.

While traces of our atmosphere can be found as far away as 600 miles from the Earth, three-quarters of its total content—our life-giving shell—is squashed down into a layer only about 6 miles thick. Mount Everest sticks out to the atmosphere's edge; the remaining 594 miles will not support life.

The Origin of Our Life-Giving Blanket of Air

Our blanket of air is a mixture of constantly churning, highly compressed gases. This churning action is what provides us with everything from those perfect, beautiful spring days to the most destructive violent storms. In the macro sense, compared with the atmosphere of other planets in our solar system, the Earth's atmosphere is calm and benign. Yet in the micro sense—our own frame of reference—this churning action rarely seems calm and can wreak too much havoc to be called benign.

The gases that constitute our atmosphere originated from within the Earth itself. Millions of years ago, when the planet was still in its violent developmental stages, countless volcanoes belched out huge quantities of poisonous gases that were generated by the chemical reactions under the Earth's crust.

It may not seem to make sense that primitive plants survived in this layer of boiling chemicals, but plants can survive in conditions that will not support animal life. Plants breathe carbon dioxide and flourish in bright sunlight. Since carbon dioxide is a by-product of the combustion processes that formed our planet, plenty was available for plant life. Also, the atmosphere millions of years ago was very thin, permitting more sunlight to reach Earth's surface. These conditions enabled early plant life to thrive.

As a by-product of their growth, plants break free the oxygen molecules in carbon dioxide. Thus, our atmosphere is in part the waste product of plants that lived on the Earth millions of years ago!

Today Earth's atmosphere consists mostly of nitrogen—about four-fifths of its total volume. The rest is mostly oxygen, with enough carbon dioxide to support today's plant life; it also contains traces of water vapor and other gases. The atmospheric layer does much more than provide the gases that sustain life on Earth. It also acts as an insulator, moderating the Earth's temperature to keep us warm at night and protect us from burning up during the day. It blocks most of the sun's dangerous ultraviolet-light rays, which would be too intense for life as we know it without this atmospheric filter.

What Is Weather?

Our atmosphere supplies one other element essential to all life on Earth: water, in the form of vapor. It distributes water vapor because it is constantly in motion, stirring and mixing its elements. This motion is caused by two factors: the uneven heating of the Earth's surface (uneven because the sun shines on only half the planet at a time) and the Earth's rotation. As atmospheric gases expand and contract, through heating and cooling, this rotation stirs the elements and creates a constant churning.

Warm air expands, becoming lighter than cool air, and rises away from the Earth's surface. Cold air condenses, becomes heavier, and sinks down. Therefore, sunlight entering the cool polar regions passes through a thicker layer of air than does sunlight entering at the warm equator. The result is hotter air at the equator and colder air near the poles. Equatorial air rises, producing low pressure; polar air sinks, producing high pressure.

The laws of equilibrium constantly try to even out the differences in temperature and pressure caused by the sun's uneven heating of the atmosphere. Thus, equatorial air rises and flows toward the Earth's poles. Cold polar air is drawn back toward the equator, and the Earth's rotation puts a spin on the whole operation. The result is a global churning of the atmosphere.

This churning action and the associated temperature and pressure differences have dramatic effects on the water vapor in the atmosphere, turning it into rain, snow, sleet, and hail, which fall to Earth as our source of life-giving water. These phenomena, which we call *weather,* are the visible results of our atmosphere's effect on the water vapor it carries.

All the energy that drives the churning action comes from our sun's heat and powers the Earth's weather system. The laws of equilibrium cannot keep up with the enormous amount of energy that the sun radiates into our atmosphere. The result is an uneven distribution of temperature, air pressure, and water content across the planet. Now we'll take a look at how varied these conditions are.

Global Climate-Variation Charts

Around the globe, the uneven distribution of temperature, moisture, and pressure combines with geographical differences to create a vast array of variations in the weather. These differences are grouped, in broad terms, into *climates.* The distinction from one climate to another may be blurred by wide zones of gradual change, but overall the groupings represent clear differences.

The main criteria for distinguishing among climates are the average annual temperature and rainfall of the region. Some classification systems also consider such factors as humidity and wind. The climates discussed here are based on the work of Vladimir Köppen, published in 1900. Köppen's system is based on the correlation between temperature and rainfall averages with the vegetation they support.

Climate types are not a reliable guide to predicting an area's weather unless local conditions are relatively constant. Changing local conditions can produce great variations in local weather. Furthermore, the Earth's climates are not static. Over long periods of time, climates change. There is some evidence that humans may be speeding climatic change through our effects on the Earth's atmosphere, though many scientists disagree with this conclusion. Those who believe humans are changing the Earth's climate attribute the cause to the *greenhouse effect,* a topic that is discussed later in this chapter.

A couple of examples of how the greenhouse effect is believed to have affected climates are the

CHAPTER 1: HISTORY

gradual desertification of the Sahel and unusual weather in Britain.

The Sahel, lying just south of the Sahara in Africa, experienced severe drought in the 1980s that threatened with famine more than 35 million people across Senegal, Mali, Niger, Nigeria, Chad, Burkina Faso, and Sudan. These areas experienced a drop in water levels of 40 to 90 percent, which dried up streams and wells and killed off vast areas of grasslands.

Britain, never known for heat waves, recently has experienced an unprecedented series of long, hot summers. It even had a hurricane in 1987.

Some experts now believe, however, that the greenhouse effect is overrated and that these variations may be more natural than we assumed earlier. Later in this chapter, after the greenhouse-effect discussion, we'll present an update on recent studies of the greenhouse effect that offer a different—and more upbeat—opinion than we've been hearing the last few years.

In the recent past Earth has been divided into five basic climatic regions:

- mild humid climates
- cold humid and cold polar climates
- tropical rainy climates
- dry climates
- mountain climates

Mild Humid Climates

Most of the world's population live in the mild humid climatic regions. There is a wide range of differences among the areas grouped under this classification.

For example, all the areas with Mediterranean-type climates are similar in their seasonal rainfall records, having hot, dry summers with prevailing winds blowing from over land and milder, wetter winters. Those mild humid climates located in upper latitudes generally are more changeable. The more southerly areas generally are warmer with heavy winter rains, possibly including monsoons.

Mild Humid Climates
Monthly Average Temperatures (in Degrees Fahrenheit) and Rainfall (in Inches)

Place	J	F	M	A	M	J	J	A	S	O	N	D
Adelaide, Australia	74°	74°	70°	64°	58°	54°	52°	54°	57°	62°	67°	71°
	0.7	0.7	0.1	1.8	2.8	3.1	2.7	2.5	2.0	1.7	1.2	1.0
Athens, Greece	48°	49°	52°	59°	66°	74°	82°	82°	73°	66°	57°	52°
	2.0	1.7	1.2	0.9	0.8	0.7	0.3	0.5	0.6	1.6	2.6	2.6
Cape Town, South Africa	70°	70°	68°	63°	59°	56°	55°	56°	52°	61°	64°	68°
	0.7	0.6	0.9	1.9	3.8	4.5	3.7	3.4	2.3	1.6	1.1	0.8
Chungking, China	48°	50°	58°	68°	74°	80°	83°	86°	77°	68°	59°	50°
	0.7	0.9	1.3	4.0	5.3	6.7	5.3	4.4	5.8	4.6	2.0	0.9
Dunedin, New Zealand	58°	58°	55°	52°	47°	44°	42°	44°	48°	51°	53°	56°
	3.4	2.7	3.0	2.7	3.2	3.2	3.0	3.1	2.8	3.0	3.3	3.5
Lisbon, Portugal	51°	52°	54°	58°	60°	67°	70°	71°	68°	62°	57°	52°
	3.6	3.5	3.4	2.6	2.0	0.8	0.2	0.2	1.4	3.3	4.3	4.1

Place	J	F	M	A	M	J	J	A	S	O	N	D
Paris, France	37° 1.5	39° 1.4	43° 1.6	49° 1.7	56° 1.9	62° 2.1	65° 2.2	64° 2.1	58° 1.9	50° 2.3	43° 1.9	38° 2.0
New Orleans, United States	54° 4.5	57° 4.3	53° 4.6	69° 4.5	75° 4.1	80° 5.4	82° 6.5	81° 5.7	78° 4.5	69° 3.2	61° 3.8	55° 4.5
San Francisco, United States	49° 4.8	51° 3.6	53° 3.1	54° 1.0	56° 0.7	57° 0.1	57° 0.0	58° 0.0	60° 0.3	59° 1.0	56° 2.4	51° 4.6
Washington, D.C., United States	34° 3.2	35° 3.0	43° 3.5	54° 3.3	64° 3.6	72° 3.9	77° 4.4	74° 4.0	68° 3.1	57° 3.1	46° 2.5	36° 3.1

Cold Humid and Cold Polar Climates

From the following table you can spot two distinct variations within this climate grouping. One group of areas has no dry season, is influenced by moist westerly winds, and lies on the western side of continents. As you move eastward across the continents on which these areas lie, you'll see greater variation in annual rainfall and temperature.

Areas in the other group have a distinct dry season and lie on the eastern side of a large continent. These are found only in Eurasia, however, because the continents in the Southern Hemisphere do not lie close enough to the South Pole to experience such cold climates.

Cold Humid and Cold Polar Climates

Monthly Average Temperatures (in Degrees Fahrenheit) and Rainfall (in Inches)

Place	J	F	M	A	M	J	J	A	S	O	N	D
Barnaul, Russia	0° 0.8	3° 0.6	14° 0.6	34° 0.6	52° 1.3	63° 1.7	68° 2.2	62° 1.8	51° 1.1	35° 1.3	17° 1.1	6° 1.1
Dawson, Canada	-23° 0.8	-11° 0.8	4° 0.5	29° 0.7	46° 0.9	57° 1.3	59° 1.6	54° 1.6	42° 1.7	25° 1.3	1° 1.3	-13° 1.1
Harbin, China	-2° 0.1	5° 0.2	24° 0.4	42° 0.9	56° 1.7	66° 3.8	72° 4.5	69° 4.1	58° 1.8	40° 1.3	21° 0.3	3° 0.2
Helsinki, Finland	21° 1.8	20° 1.4	25° 1.4	34° 1.4	46° 1.8	57° 1.8	62° 2.2	60° 2.9	52° 2.5	42° 2.6	32° 2.5	25° 2.4
Kazan, Russia	7° 0.5	10° 0.4	20° 0.6	38° 0.9	54° 1.6	63° 2.2	68° 2.4	65° 2.4	51° 1.6	39° 1.1	25° 1.0	11° 1.7
Montreal, Canada	13° 3.7	15° 3.2	25° 3.7	41° 2.4	55° 3.1	65° 3.5	69° 3.8	67° 3.7	59° 3.5	47° 3.3	33° 3.4	19° 3.7

Place	J	F	M	A	M	J	J	A	S	O	N	D
Moscow, Russia	12° 0.9	15° 0.7	24° 0.8	38° 0.7	53° 1.2	62° 1.8	66° 2.4	63° 2.4	52° 2.2	40° 1.6	28° 1.2	17° 0.9
Winnipeg, Canada	-4° 0.9	0° 0.7	15° 1.2	38° 1.4	52° 2.0	62° 3.1	66° 3.1	64° 2.2	54° 2.2	41° 1.4	21° 1.1	6° 0.9
Yakutsk, Russia	-46° 0.9	-35° 0.2	-10° 0.4	16° 0.6	41° 1.1	59° 2.1	66° 1.7	60° 2.6	42° 1.2	16° 1.4	-21° 0.6	-41° 0.2

Tropical Rainy Climates

Within this general climatic grouping, you'll find three types: equatorial climates, tropical climates, and monsoon climates.

The equatorial climates lie within the area of convergence of the Northeast and Southeast trade winds—steady winds that trade ships relied upon in the days when wind was their primary source of power. These areas are characterized by heavy annual rainfall (they have no dry season) and by high temperatures. The rainfall is due to the great vertical motion of moist air within those regions, while their high temperatures are due to their latitude.

Like the equatorial climates, the tropical climates have heavy rainfall and high temperatures, but they also experience a short dry season. The slight seasonal nature of these areas occurs because the trade winds shift somewhat, so the areas lie under the zone of convergence of the trade winds, thus causing a dry season, or at least a lessening of the rains.

The monsoon areas also are tropical but experience longer dry seasons because they are influenced by the trade winds more than are the other two tropical regions. Those areas lying on the western side of a landmass may experience long dry seasons. The monsoon occurs when the Southeast trade winds are diverted from south of the equator toward India by a deep low-pressure area in Asia in the summer.

Tropical Rainy Climates
Monthly Average Temperatures (in Degrees Fahrenheit) and Rainfall (in Inches)

Place	J	F	M	A	M	J	J	A	S	O	N	D
Banjul, Gambia	74° 0.0	75° 0.0	76° 0.0	76° 0.0	77° 0.2	80° 2.9	80° 10.9	79° 19.6	80° 10.0	81° 3.7	79° 0.2	75° 0.1
Bombay, India	76° 0.1	76° 0.1	80° 0.0	83° 0.0	86° 0.7	84° 19.9	81° 24.0	81° 14.5	81° 10.6	82° 1.9	81° 0.4	77° 0.0
Calcutta, India	67° 0.4	71° 1.0	80° 1.4	85° 2.2	86° 5.6	85° 11.9	84° 12.7	83° 13.4	83° 10.0	81° 4.9	73° 0.6	67° 0.2
Caracas, Venezuela	65° 0.9	65° 0.3	66° 0.6	68° 1.6	70° 2.8	69° 4.0	68° 4.3	68° 4.2	69° 4.1	68° 3.8	67° 3.3	65° 1.8
Colón, Panama	80° 3.7	80° 1.6	80° 1.6	81° 4.3	81° 12.4	80° 13.3	80° 16.0	80° 14.8	80° 12.5	80° 15.1	80° 20.7	79° 11.4

Place	J	F	M	A	M	J	J	A	S	O	N	D
Jakarta, Indonesia	79° 13.0	79° 12.8	80° 7.8	81° 5.1	81° 4.0	80° 3.7	80° 2.6	80° 1.7	81° 2.9	81° 4.5	79° 5.5	79° 8.5
Kananga, Zaire	76° 7.2	76° 5.4	76° 7.9	77° 6.1	77° 3.1	76° 0.2	77° 0.1	76° 2.5	76° 6.5	76° 6.6	76° 9.1	77° 6.6
Manaus, Brazil	80° 9.2	80° 9.0	80° 9.6	80° 8.5	80° 7.0	80° 3.6	80° 2.2	81° 1.4	82° 2.0	83° 4.1	83° 5.5	82° 7.7
Manila, Philippines	77° 0.8	78° 0.4	80° 0.8	83° 1.3	83° 4.5	82° 9.2	81° 17.3	81° 16.0	80° 14.3	80° 6.7	78° 5.2	77° 3.1
Nairobi, Kenya	64° 1.9	65° 4.2	66° 3.7	65° 8.3	63° 5.2	61° 2.0	59° 0.8	60° 0.9	63° 0.9	66° 2.0	64° 5.8	63° 3.5
Rangoon, Myanmar	77° 0.2	79° 0.2	84° 0.3	87° 1.4	84° 12.1	81° 18.4	80° 21.5	80° 19.7	81° 15.4	82° 7.3	80° 2.8	77° 0.3
Yaoundé, Cameroon	74° 1.6	74° 2.7	74° 5.9	72° 9.1	72° 8.1	71° 4.5	70° 2.6	71° 3.3	71° 7.6	71° 8.9	72° 5.9	73° 2.0

Dry Climates

This climate classification has two distinct types, desert and semiarid. Normally, we associate a desert climate with extremes of heat, such as occur in the Sahara Desert. Deserts, however, result from a lack of rain. Walvis Bay, South Africa, is an excellent example, with virtually no annual rainfall yet with moderate average temperatures. The distinction between the two types, desert and semiarid, is based on more than the simple measure of total rainfall. It also must include consideration of the effectiveness of the precipitation.

Such scant precipitation as these areas receive is useless for growing crops if it is spread evenly through the year or if it comes in sporadic and unpredictable rains. The dividing line between desert and semiarid climates usually is about 15 inches of annual rain. Yet an area can receive substantially less rainfall than that and still support crop growth if the timing is right and the rain is reliable. A good example is the western part of Australia, which receives only 10 inches per year, but the rain is reliable and timed perfectly for crops.

Deserts are created when their positions relative to their dominant high-pressure systems cause them to get air masses that are mostly dry. You might expect an area located on the western coasts of continents, such as the Namib Desert, to get plenty of rain from their proximity to the ocean. The prevailing winds over the Namib, however, arrive from over land and bring dry air and therefore little to no rain.

Dry Climates

Monthly Average Temperatures (in Degrees Fahrenheit) and Rainfall (in Inches)

Place	J	F	M	A	M	J	J	A	S	O	N	D
Aden, Yemen		76° 0.3	77° 0.2	79° 0.5	83° 0.2	87° 0.1	89° 0.1	88° 0.0	87° 0.1	88° 0.1	84° 0.1	80° 0.1	72° 0.1

CHAPTER 1: HISTORY

Place	J	F	M	A	M	J	J	A	S	O	N	D
Alice Springs, Australia	84°	82°	77°	68°	60°	54°	52°	58°	66°	74°	80°	82°
	1.8	1.7	1.3	0.9	0.6	0.6	0.4	0.4	0.4	0.7	0.9	1.3
Kashgar, China	22°	34°	47°	61°	70°	77°	80°	76°	69°	56°	40°	26°
	0.3	0.0	0.2	0.2	0.8	0.4	0.3	0.7	0.3	0.0	0.0	0.2
San Diego, United States	54°	55°	57°	58°	61°	64°	67°	68°	67°	63°	59°	56°
	1.8	1.9	1.5	0.6	0.3	0.1	0.1	0.1	0.1	0.4	0.9	1.8
Tarfaya, Morocco	61°	61°	63°	64°	65°	67°	68°	68°	69°	68°	65°	62°
	0.5	0.5	0.5	0.0	0.0	0.0	0.0	0.5	0.5	0.5	0.5	1.0
Tehran, Iran	34°	52°	48°	61°	71°	80°	82°	83°	77°	66°	51°	42°
	1.6	1.0	1.9	1.4	0.5	0.1	0.2	0.0	0.1	0.3	1.0	1.3
Walvis Bay, South Africa	65°	66°	66°	65°	62°	60°	59°	57°	58°	60°	61°	64°
	0.0	0.0	0.0	0.0	0.0	0.0	0.0	0.0	0.0	0.0	0.0	0.0

Mountain Climates

Local weather within a generally mountainous region can vary greatly due to differences in elevation. Valleys may receive very low total rainfall, while a nearby mountain range gets high rainfall. Some are so high, as in the High Andes and the highest areas of the Himalayas, that their climates most resemble a polar type. The Quito Valley in Ecuador lies almost on the equator, where one might expect high temperatures to prevail, yet it has a perpetually spring-like climate due to its 9,400-foot altitude.

Most mountain climates are more typical of their surrounding regions, such as the lower Andes, where the expected tropical climate is modified by the mountain elevations. Still, within even the normal mountain-type climate, you'll find great variations. The Alps are a modified cool temperature climate. The foothills of the Himalayas have a highly modified mild humid climate to the south. The plateau of Tibet gets a modified version of the arid climate to its north.

Mountain Climates

Monthly Average Temperatures (in Degrees Fahrenheit) and Rainfall (in Inches)

Place	J	F	M	A	M	J	J	A	S	O	N	D
Arequipa, Peru	58°	58°	58°	58°	58°	57°	57°	57°	58°	58°	58°	58°
	1.2	1.7	0.6	0.2	0.0	0.0	0.0	0.0	0.0	0.0	0.1	0.4
Davos, Switzerland	19°	23°	27°	36°	41°	50°	54°	52°	47°	38°	30°	21°
	1.8	2.2	2.2	2.2	2.3	4.0	4.9	5.0	3.7	2.7	2.2	2.5
Leh, India	17°	19°	31°	43°	50°	52°	63°	61°	54°	43°	32°	22°
	0.4	0.3	0.3	0.2	0.2	0.2	0.5	0.5	0.3	0.2	0.0	0.2

Still Constantly Changing

The climate patterns described in this chapter have not been around forever. Our atmosphere has evolved greatly from its composition of millions of years ago. Today, on our micro-scale frame of reference, it appears stable, yet the Earth is still changing. While volcanoes continue to affect the atmosphere, another factor is at play that was not present millions of years ago: the human impact. First let's discuss how the ancient volcano factor continues to alter climate and weather, then we'll take up the subject of how humans have become agents of climatic change. We'll wrap up with a prediction of how human activities will influence the future of global climates.

The Earth Isn't Done Yet: Volcanoes

Explosive volcanic eruptions emit huge quantities of gases and fine debris into the atmosphere and may cause some short-term climatic variability, but the long-term impact on climate of a single volcanic eruption, no matter how great, is relatively small. The ejected volcanic material is projected high into the stratosphere, where it circles around the globe and remains for many months. It is believed to filter out part of the incoming solar radiation, thereby lowering air temperature.

Many scientists agree that the impact of a volcanic eruption on global temperatures is minor but that the cooling that is produced may alter the pattern of atmospheric circulation for a short time, causing some regional weather changes. For example, when Mount Tambora in Indonesia erupted in 1815, the period that followed was termed "the year without a summer"—the unusually cold temperatures in the Northern Hemisphere during 1816 were attributed to the cloud of volcanic debris emitted from Tambora.

Atmospheric scientists, however, are not able to predict and identify these short-term regional effects reliably, and recent studies have not found any link between volcanoes and long-term climatic changes. When Mount St. Helens erupted in 1980 and El Chichón in Mexico erupted in 1982, scientists were able for the first time to monitor the atmospheric effects of volcanic eruptions with sophisticated technology, including satellite images and remote sensing instruments. They found that the large quantity of volcanic ash emitted by the Mount St. Helens eruptions did have immediate regional effects but that the cooling was so minimal it could not be distinguished from natural temperature fluctuations.

Explosiveness alone is a poor criterion for predicting the global atmospheric effects of an eruption. This was determined after comparing the El Chichón eruption to that of Mount St. Helens. The Mount St. Helens eruption was more explosive, but the material emitted was largely fine ash, which settled in a short time. El Chichón emitted an estimated forty times more sulfur-rich gases, which combined with water vapor in the stratosphere to produce a dense cloud of tiny sulfuric-acid droplets. Such clouds take years to settle completely and are capable of decreasing the mean global temperature because the droplets absorb solar radiation and scatter it back to space. Long-lived volcanic clouds are not composed of dust, as was once thought, but consist largely of sulfuric-acid droplets. The Mount St. Helens eruption thus did not have the gases needed for a long-term effect.

Many great volcanic eruptions—occurring closely together—must have a pronounced impact on climate. No such period of explosive volcanism is known to have taken place in historic times. Still, other geological theories exist to explain global climatic changes. One is the plate-tectonics theory.

The Plate-Tectonics Theory of Climatic Changes

The plate-tectonics theory, which has emerged over the last few decades and is widely accepted by scientists, states that the outer part of the Earth is made of many large separate plates that float on a partially molten layer of the Earth's crust and move in relation to one another. The plates consist of both continental and oceanic crust, and the continents shift positions as the plates move. This theory gives

The Eruption of Mount St. Helens in 1980

both geologists and climatologists a probable solution for some previously unexplained climate changes (although it is not useful for explaining short-term changes).

After the plate-tectonics theory was formulated and proven, scientists realized that several present-day continents were once joined together as one "supercontinent." This grand landmass broke apart, and its pieces—each moving on a different plate—slowly moved to their current locations. This theory thus explains the ancient glacial features found in the scattered tropical regions of today's Africa and South America.

Landmasses shifting to different latitudinal positions accounted for many dramatic climate changes. (Scientists believe that the transport of heat and moisture in oceanic circulation must also have changed, altering climates as well.) Changes in the positions of continents take millions of years because the rate of plate movement is very slow—only a few centimeters per year. Climatic changes brought about by continental drift are very gradual, which is why the plate-tectonics theory is not helpful for explaining the climatic variations occurring during shorter time periods.

The Greenhouse Effect

There is little basis for expecting the Earth to undo its own climate, but what about its inhabitants? Is humankind going to cause the Earth's climate to become uninhabitable? Many scientists and environmentalists predict exactly that. The vehicle for this artificially induced destruction, they claim, will be the greenhouse effect.

In an actual greenhouse, such as those used at plant nurseries, sunlight penetrating the glass panels heats the inside; the glass keeps the resulting heated air trapped inside. You've seen the greenhouse effect at work if you've ever opened a locked car that has been parked in direct sunlight.

When discussions of climate include mention of the greenhouse effect, they are referring to a phenomenon in which certain gases in the Earth's atmosphere—mainly carbon dioxide—act like the glass panels in a nursery greenhouse. The gases create a layer that lets sunlight penetrate to heat the air below but then traps the heat and keeps it from radiating back into space.

This atmospheric greenhouse phenomenon creates a couple of carryover effects that amplify the basic greenhouse action. First, since the trapped heat does not rise (as normally would be expected of hot air), the air in the upper atmosphere turns cooler than normal. This cooling means that the water vapor in the upper atmosphere is more likely to condense into clouds; this in turn increases the effects of the heat trap caused by the carbon dioxide layer. Second, the warming climate that results from this trapped heat causes more water to evaporate, thus increasing the cloud cover further and leading to even more trapped heat.

According to a study by the U.S. Department of Energy, cloud cover over the United States increased by 3.5 percent between 1950 and 1988. In 1988 another study by the Department of Energy concluded from numerous shipboard weather observations that the skies over the oceans also have grown cloudier. A study by the German government reported a decline in sunshine in that country, as well.

In 1988 scientist James Hansen, of the Goddard Institute for Space Studies in New York City, first sounded the alarm over global warming. Since then some reports have argued that global warming has been restricted to increased nighttime temperatures and that, in some cases, daytime temperatures actually have decreased.

In an updated report on global warming, Hansen disputes the rosier predictions, claiming that recent increases in cloud cover support his earlier warnings about the greenhouse effect. He thinks that the increased cloudiness of today is caused by particles of pollution in the atmosphere—mostly sulfates from the burning of fossil fuels. His theory states that this increased cloud cover reduces daytime temperatures by blocking the sun during the day and that it increases nighttime temperatures by trapping warm air near the Earth's surface.

Hansen suggests that the increased cloudiness caused by sulfates will abate over time as we do a better job at controlling atmospheric sulfate emissions. Also, he notes, global consumption of fossil fuels is not increasing as fast as was earlier predicted. However, he warns, this apparent abeyance in global warming is only temporary. While the lower fossil-fuel use means less carbon dioxide in the atmosphere to cause the greenhouse effect, it may be decades before there is a significant decrease of carbon dioxide in the Earth's atmosphere. The carbon dioxide already there from previous consumption will not dissipate quickly, enabling it to contribute to the greenhouse effect long after output declines.

The sulfates that are causing today's increased cloud cover are washed from the air in days as rain falls. According to Hansen, eventually this will lead to a return to more normal cloud cover and result in increased heating by the sun. Since carbon dioxide will still be aloft in our atmosphere, he predicts that the warming factors will overtake the cooling factors and lead to a significant global temperature rise during the day as well as at night.

Update on the Greenhouse Effect

New reports based on studies of the greenhouse effect show that expected global warming has been relatively more benign than was originally predicted. The most optimistic report to be issued on the greenhouse effect comes from Patrick J. Michaels, a professor of environmental sciences at the University of Virginia and the Virginia state climatologist.

Michaels flatly states that "the popular vision of climate apocalypse is wrong." Michaels sees the most dramatic change from global warming as a

Increased Cloud Cover—A Result of Pollution?

lengthening of the growing seasons in agricultural regions, because the frost-free season would begin sooner each spring and extend later into the fall.

Michaels, who has long doubted the catastrophic predictions of global warming, bases his optimism on data that have been published many times in tables of numerical weather data. He says he has not made a new discovery but is merely pointing out the significance of something that has been reported for a long time. Michaels states of his theory, "Nobody ever pulled it together. It was there, waiting to be synthesized from the literature."

Considering Michaels's report, one might wonder about James Hansen's concerns about the effects of lingering high carbon-dioxide levels. Michaels believes that Hansen is wrong regarding both the amount of warming and the role of aerosol particles like sulfates. Hansen's computer models would predict a temperature rise about twice what has occurred over the last century. These models predicted that carbon-dioxide levels would double by the year 2030 and that the average global temperatures would be as much as 5°F higher than they are now.

In contrast, Michaels' report notes that carbon-dioxide levels are not expected to double until after the middle of the next century. Furthermore, they then are expected to plateau. Michaels also disagrees with Hansen's conclusion on how aerosol particles will affect global warming by contributing to increasing cloud cover. Michaels notes that aerosol particles do not rise to the upper atmospheric levels. Since they remain in the lower atmosphere (the "haze zone" visible from airplanes), they will not contribute to increased cloudiness, which takes place at much higher altitudes.

Other scientific data have shown that some of the warnings about global warming may be exaggerated. Two scientists, P. D. Jones and K. R. Briffa, from the climate research center at the University of East Anglia, note that the only increases in global temperatures have been in the spring, fall, and autumn. Summer temperatures, they report, are no warmer now than during the U.S. Civil War.

Jones and Briffa also find that polar temperatures are not increasing as rapidly as once was predicted by computer models. Thus, the main danger of global warming—a rise in mean sea level as the ice caps melt—does not seem to be imminent. For the ice caps to pose a serious threat to coastal regions, they would have to melt faster in summer than they grow in winter.

Further lessening the threat of dire predictions coming true is the following argument: If the poles did warm, snowfall would increase because it is now often too cold to snow there; at the same time the melting rate would increase, producing more moisture for snowfall. These two opposing trends would approximately cancel each other out.

Nonetheless, we all contribute to global pollution in some measure, and the problem won't go away unless we each contribute to solutions. We can find better ways to contribute if we understand how our presence affects the environment. Fortunately, there are now some excellent on-line resources that can help you understand the future impact of mankind on our planet's climate.

The Carbon Dioxide Information Analysis Center

You can learn more about the future health of your atmosphere from the Internet by visiting the Carbon Dioxide Information Analysis Center (CDIAC) at *http://cdiac.esd.ornl.gov*. This site also will link you to a variety of related on-line resources.

The Carbon Dioxide Information Analysis Center, which includes the World Data Center-A for Atmospheric Trace Gases, is the primary global-change data and information analysis center of the U.S. Department of Energy (DOE). More than just an archive of data sets and publications, CDIAC has—since its inception in 1982—enhanced the value of its holdings through intensive quality assurance, documentation, and integration. As opposed to many traditional data centers that are discipline-based (e.g., meteorology or oceanography), CDIAC's scope includes potentially anything and everything

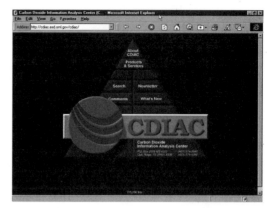

(Courtesy of the Carbon Dioxide Information Analysis Center)

that would be of value to users concerned with the greenhouse effect and global climate change, including concentrations of carbon dioxide and other radiatively active gases in the atmosphere; the role of the terrestrial biosphere and the oceans in the biogeochemical cycles of greenhouse gases; emissions of carbon dioxide to the atmosphere; long-term climate trends; the effects of elevated carbon dioxide on vegetation; and the vulnerability of coastal areas to rising sea level.

National Climatic Data Center

The National Oceanic and Atmospheric Administration (NOAA) brings you a wealth of on-line climatic information. Visit their Web site at *http://www.ncdc.noaa.gov* for details. Their home

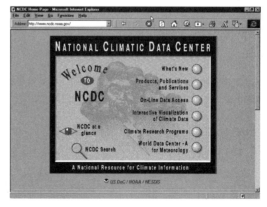

(Courtesy of the National Climatic Data Center)

page will lead you to all kinds of products, publications, and services that will help you learn more about the human impact on our atmosphere and what you can do to help. They've got great charts, graphs, and photos that you can download on-line and print anytime.

Global Weather Products

Take Charge of Climatic History—with "Sim Earth"
Maxis Software
Suite 230
2 Theatre Square
Orinda, CA 94563-3346
(510) 254-9700   fax: (510) 253-3736

Now you can understand the evolution of our planet's climatic history as never before. With "Sim Earth," a challenging, educational, exciting computer game for IBM-PC clones and Macintoshes, you can create landmasses, climates, and animal life forms as you choose. Then, using this astonishing computer simulation of the planetary life cycle, you can guide planetary development and see how it affects animal evolution. Introduce a radical temperature change or create other hostile environmental disasters and see what species survive. But don't take too long—the program only gives you 10 billion years before the heat from the sun destroys all life. This program requires a hard-disk drive and sells for $69.95.

Tracking Pollution and Global Warming
American Weather Enterprises
P.O. Box 1383
Media, PA 19063
(215) 565-1232

American Weather Enterprises sells books, software, and educational materials, including some that detail global atmospheric influences, meteorological motion, severe weather, clouds and precipitation, examples of pollution, aviation meteorology, and more in an easy-to-understand flowchart design. The company also offers Geochron, a beautiful, framed electronic map that continually shows the exact time of day and amount of sunlight anywhere in the world. It will help you understand our planet's climates and weather patterns.

World Climates
C. W. Hickox
Program in the GeoSciences
Emory University
Atlanta, GA 30322

Using the Vladimir Köppen climate classifications, Hickox has made a complete set of world climate data available for Macintosh users via the HyperCard. Climates can be listed, displayed graphically, and superimposed on one another. You can choose cities or whole continents, even on opposite sides of the planet. You can export the climate data to a spreadsheet or database program.

Climate Data
The sources listed below have meteorological, oceanographic, and geophysical data available on tape and other electronic media. Some of the data types that they maintain are intended only for hobbyist use; other data are more research-oriented. Much of the research data requires a fee payment.

Carbon Dioxide Information Analysis Center (CDIAC)
Oak Ridge National Laboratory
P.O. Box 2008
Oak Ridge, TN 37831-6335
(615) 574-0390
fax: (615) 574-2232
e-mail: cdp@stc10.ctd.ornl.gov

Langley Distributed Active Archive Center (DAAC)
NASA Langley Research Center
MS 157B
Hampton, VA 23681-0001
(804) 864-8656
fax: (804) 864-8807
e-mail: userserv@eosdis.larc.nasa.gov.

CHAPTER 1: HISTORY

NASA Space Science Data Center (NSSDC)
NASA Goddard Space Flight Center
Greenbelt, MD 20771
(301) 286-6695

National Center for Atmospheric Research (NCAR)
NCAR Data Support Section
P.O. Box 3000
Boulder, CO 80307
(303) 497-1219
fax: (303) 497-1137
e-mail: datahelp@ncar.ucar.edu

National Climatic Data Center (NCDC)
Federal Building
Asheville, NC 28801
(704) 259-0682

United Nations Environment Program (UNEP)
Global Resource Information Database (GRID)
UNEP/GRID-Geneva
Facility Manager
6 rue de la Gabelle, CH-1227
Carouge, Geneva, Switzerland
(0041-22) 343-8660
fax: (0041-22) 343-8862
e-mail: POSTMAN@grid.unige.ch

UNEP/GRID—Nairobi (For Africa and Latin America)
Facility Manager
P.O. Box 30552
Nairobi, Kenya
(00254-2) 230-800
fax: (00254-2) 226-491

Chapter 2
Weather Lore

Nothing in nature has spawned more folklore than our favorite topic: the weather. For centuries mankind's insatiable desire to predict the weather has led people to turn to some amazingly bizarre "meteorologists," so you might expect this chapter to relate ancient lore that arose out of some primitive civilizations. Well, hang on to your expectations. The stories are not all ancient. Even in the United States, with its armada of orbiting weather satellites and army of Cray supercomputers crunching billions and billions of bytes, people still believe in some down-to-earth tales.

Goats Tell It on the Mountain

It might have been an over-lively meteorological debate that made two local forecasters burst into Beverly Bouska's garden and begin brawling on her sun deck. Before this donnybrook was over, the combatants destroyed a greenhouse, a maple tree, and a fence. It was this escapade—the last among numerous difficulties with her pugnacious neighbors—that finally got Ms. Bouska's goat; she put her house on the market.

Who were these unwelcome and unruly prognosticators? They were, in fact, a pair of the legendary weather-forecasting goats of Mt. Nebo, near Roseburg, Oregon. Such was the herd's fame that some twenty years ago they were featured on the NBC Nightly News with David Brinkley and appeared in magazines and newspapers worldwide.

The goats' meteoric (as well as barometric) rise to fame began in May 1971, when Tom Warden of Roseburg's radio station KRSB noticed it was raining hard outside, although the Weather Bureau's forecast had called for a cloudy morning followed by afternoon sunshine. He then recalled the Roseburg folklore concerning the goats, a handful of which had been grazing the 1,200-foot-high hogback mountain by Interstate 5 for years.

Locals reportedly said that when the goats were up the mountain, the weather would be dry or fair, while the reverse applied when they grazed lower down on the slopes. Local pilots were said to use the goats as a weather check and housewives consulted the goats before hanging the laundry outside. Warden was inspired to compare the accuracy of the goats' and the Bureau's forecasts for a week. The result? The animals achieved 90 percent accuracy, while the Bureau was right only 65 percent of the time.

Warden and KRSB began giving daily Goat Weather Forecasts, using such shorthand as "widely scattered goats" for sunshine and "low goat pressure" to mean storms, rain or snow on the way. These forecasts became so popular that a Goat Observation Corps formed, complete with t-shirts and membership cards promoting the "World's Only Weathervane Goats." The horned forecasters were local heroes: a 1978 Roseburg map proudly proclaims, "Home of the Famous Weather Forecasting Goats," and a sweatshirt declaring "Ski Mount Nebo" depicts the goats hitting the slopes.

Local historian Homer Brown's *In Search of the Mount Nebo Goats,* a manuscript at the Douglas County (Oregon) Museum of History and Natural History, tells of a local prankster who cut out cardboard goats and erected them on the mountainside one night. The next morning Tom Warden, following the traditional forecasting method, checked the goats on Mt. Nebo and naturally based his Goat Weather Forecast on what he saw, with predictably wayward results. The gag was up when Warden noticed before giving the next report that the goats had not moved all morning. Apparently the cardboard was traced to the culprit, who was arrested for littering, taken to the police station in handcuffs, and locked up until he paid a fine. He may be the only person in history to be arrested for goat littering.

After the Oregon State Highway Department ironed out a curve on I–5 near the base of the mountain, the goats began foraging further afield, including on or near the road itself. Black and yellow warning signs advising motorists of the "Goat Crossing" were added. More serious problems emerged in the spring of 1978, when youths shot and killed two of the goats (taking a head as a trophy) and seriously wounded a third. Fortunately it was nursed to health and returned to the herd.

There were other concerns, as well. Roseburg's local paper, the *News-Review,* reported in August the following year that a concerned citizen had tipped off the sheriff's department to a literal case of kidnapping. The alleged abductor had driven off with a kid in the back of a station wagon. The accused cleared his name by revealing that he had been trying to save the animal. Several goats had been killed over the years by wandering into the paths of cars.

Clearly, the goats were a danger to themselves and others. In 1979, after Ms. Bouska's unfortunate experience, the district attorney, William Lasswell, called for the removal of the "criminal goats" in the herd. Nonetheless, the herd flourished on Mt. Nebo until fairly recently, when the problem of wandering goats became so severe that all were rounded up and shipped off to a farm near Oakland, Oregon, about 15 miles from the scene of their meteorological triumphs, thus closing a unique chapter in the history of weather forecasting and insurance claims.

—By Mary Reed. Reprinted with permission from *Weatherwise.*

Wild weather superstitions are not restricted to wild animals. The most domestic of all animals—common house cats—also are highly revered for their uncanny weather-forecasting abilities. The following story may help you realize you already have a home weather station that works purr-fectly.

Feline Forecasters

People from many different climes and times—some as well-known as Benjamin Franklin, other as obscure as a Puritan New England housewife—have relied on their cats as meteorological almanacs.

Our colonial ancestors considered their cats among the best four-footed weather forecasters in the early American menagerie of pets and livestock. Cats often were given serious treatment in almanacs. Farmers, fishermen, housewives and sailors all looked to their felines for vital information about the weather and the future.

CHAPTER 2: WEATHER LORE

Though much of what our forebears believed since has been debunked as myth or superstition, the cat as forecaster of weather may hold some credibility. As sailor and folklorist Horace Beck says:

> Much weather knowledge of a short-term nature can be learned by observing animals, and no one has yet come to any certain conclusions as to whether or not animal behavior, accidental or intentional, reflects future seasons.

The cat's sensitivity to even the most minute physical changes in its surroundings has long been regarded with suspicion and awe. It is likely that those super-tuned feline perceptions contributed to the cat's reputation as a familiar for witches and demons. The mere fact that a cat's senses so far exceed ours makes the animal seem unusual. But it may also make the cat a reliable barometer of weather changes.

I should hasten to add that most feline weather folklore has little or no sound basis. For example, there is a popular Scottish notion that, when a cat sneezes, rain is due. A number of cultures, particularly in the British Isles and Colonial America, believed that rain could be divined by observing how a cat washed itself. My Irish mother always said that a cat washing its face meant rain. Of course, given the frequency of rain in Ireland, that probably wasn't a difficult forecast.

On the other hand, an old English proverb notes:

> If a cat washes her face over her ear,
> 'Tis a sign the weather will be fine and clear.

Icelandic peoples don't care much about the cat's grooming habits, but they believe (or used to) that the position in which a cat lies down to sleep is an important weather indicator. If a cat stretches out with its paws in front of it, *a la* the Egyptian Sphinx, exceptionally foul weather is coming. As any cat lover knows, cats very rarely sleep in this position—which is why, Icelanders say, bad storms are rare.

A myth popular in many parts of the world is that a cat eating grass portends rain and storm.

In Indonesia it is believed that rain can be induced during droughts by various cat-related rituals. People of the Celebes sometimes confine a cat to a small chair and carry it over the dry fields, dousing it with water, in an effort to break a drought.

Sumatrans will toss a cat into a river, believing that rain will follow the cat as it swims ashore. In Malaysia, rain is thought to be induced by bringing a cat cruelly close to drowning. All in all, cats are probably a lot happier during rainy years in these areas.

The cat is also thought to be connected to the wind, according to folklore and religious beliefs in many lands. The ancient Japanese, for example, worshipped a wind god who took the form of a cat whose huge claws ripped open the sky before every storm.

In the Shetland Islands, a cat was thought to foretell a windy day by turning its sensitive face toward a breeze. If a cat was found asleep with the back of its head against the floor, or "sleepin' up her harns" as the Shetlanders put it, a calm day was ahead.

Sailors have always observed feline behavior with a keen eye. Cats on board ships not only eradicated rats and amused the crew, but were also regarded as being able to predict—and even control—the weather. If a cat was seen chasing its tail or playing with a dangling rope, seamen thought it was stirring up a wind. Some sailors believed a cat carried a *"gale* in its tail" and could release it at will. Hence seamen were very careful not to harm a cat or kill it, lest the gale escape.

Extremely superstitious seamen considered cats bad luck, believing they were the familiars of witches who surreptitiously came aboard and caused storms at sea. When sailors decided that their ship's cat was a "Jonah" that had brought it extremely bad luck at sea, they would toss it overboard lest it bring disaster to the vessel.

Some poor seagoing felines were called "firetail cats" because, it was thought, they were so overwrought with evil spirits that sparks flew from their tails. This was probably the result of some natural phenomenon such as static electricity in their fur, but it terrified seamen. Manx cats were popular with these crews, for a very obvious reason; since Manx have no tails, they can carry neither gale nor fire.

Sparks flying from a cat's fur, especially when it is stroked, are not unusual. This static electricity is strongest when the air is dry. New England housewives believed that if a cat's fur was rubbed the wrong way and sparks flew, it was a sign of cold, dry weather ahead. Since that is just the type of weather that promotes static electricity, they probably were correct in their belief. How the cats felt about having their fur rubbed the wrong way apparently did not matter to these amateur weather forecasters.

In the Caribbean Islands cats were thought to foretell hurricanes by nervous behavior, tail-twitching, and general restlessness. In fact a cat's keen sense of hearing may allow it to perceive the first high-pitched winds of an approaching hurricane, or the light squeals of a storm-induced wind as it sneaks through cracks in windows and doors. One of my cats, who endured a New England hurricane with me in 1985, took refuge in a box in a closet during the worst of the storm, apparently preferring to get as far as possible from the sounds of the tremendous gusts of wind.

The lore of Colonial New England relates that the cat can forecast the direction of the wind by always curling up with its back toward the compass point where the wind will originate.

A variation on this is the old wives' tale that says the cat who sits on an autumn hearth with his back facing north is prognosticating a long, cold winter.

While the cat may not be able to forecast the weather of the coming season, animals are known to place their backsides to the cold wind to protect themselves. Since most old homestead hearths were built on the coldest side of the house—the north—cats would be likely to keep their backs toward the drafty opening of the unlit fireplace. This bit of feline trivia may help explain how the "compass cat" came about.

The Puritans brought with them from England the term "a cat's nose" for a cold northwest wind, as well as a variety of other feline terms to describe the way the wind behaves. "Cat's paws" are tiny ripples on the surface of a body of water. Bigger ripples are called "cat's skin." Those small puffs of breeze that barely move the leaves on trees and merely tease the smoke rising from a chimney were called "cat's whiskers." And let's not forget my family's favorite to describe those breezes that tickle your face: "kitten fur."

On a chilly morning in late autumn, "cat ice" is predictable for puddles and the edges of small ponds. These are the fragile places where water has receded from under the ice. It looks so thin and clear that only a cat could walk on it without it breaking.

The Iroquois Indians measured the severity of a snowstorm by the tracks left by animals. Knowing that graceful predator cats take light, careful steps, the Iroquois distinguished between the dusting of snow and true blizzard by the visibility of cougar, lynx and bobcat tracks.

An old Indian saying goes: "If there is enough snow on the ground to track a cat, there has truly been a snowstorm."

Cats lying on cold hearths were not the only felines thought to be able to make long-term weather forecasts. New Englanders thought that cats who languished in the sun in February would be huddling behind the stove for warmth the following month: The cat's February sunbathing foretold a frigid March. They also gave us the quaint adage that March can come in like a lion or depart like a lion, depending on what its accomplice—the lamb—has in mind.

Farmers thought that insect populations could be predicted by studying the hearty barn cat. If a single flea was found on a cat in March, hundreds would be found in July. But if the cat was free of fleas in early spring, few would be bothering it come summer. From this bit of common sense they deduced the possible severity of pest infestations of locusts, mosquitoes, grasshoppers, weevils and moths.

Old fables, like old habits, die hard, and even today I find it fun to use terms like, "it's raining cats and dogs outside," or the cold wind "cuts like a cat's claw." And it's easy to note when the family cat is stretched out shamelessly on a sunny window sill, or has disappeared into a remote corner of the closet, and to deduce that fair or foul skies are ahead.

While I am not about to toss a cat overboard in an effort to ward off a storm at sea, I do think it makes sense to pay heed to the behavior of our feline pets. After all, cats' amazing abilities have enabled them to survive hard winters and dry summers, in jungles, in deserts and in the heart of modern cities. They are survivors, and to survive, a creature must possess a keen awareness of its surrounding environment. By watching the cat, we can benefit from that awareness that has brought it such success.

Of course, if cats do know what to expect from the skies, in their usual tacit, aloof way, they still are going to leave it up to us to figure it out on our own.

—By Elinor De Wire. Reprinted with permission from *Weatherwise*.

Other Animals

Fish
Fishes in general, both in salt and fresh waters, are observed to sport most and bite more eagerly before rain than at any other time.

Hedgehogs
Observe which way the hedgehog builds her nest,
To front the north or south, or east or west;
For it 'tis true what common people say,
The wind will blow the quite contrary way.

If by some secret art the hedgehog knows,
So long before, the way in which the winds will blow,
She has an art which many a person lacks,
That thinks himself fit to make our almanacks.

—*Poor Robin's Almanack*, 1733

The Truth about Groundhog Day

With today's vast array of high-tech equipment and computer-enhanced satellite weather photographs, it's hard to imagine relying on a furry animal for weather forecasts. Yet hundreds of years ago, Europeans used the behavior of hedgehogs to predict the end of winter. This strange custom was so well entrenched that when European settlers came to North America, they brought it with them.

When settlers first arrived in Canada, however, they could not find any of their spiny little weathermen. So, they sought a replacement forecaster that would help them learn when spring was coming in their new land. The groundhog seemed to be a good substitute.

According to these ancient traditions, groundhogs come out of their winter hibernation at noon on February 2. If the drowsy little creature sees its shadow, that frightening sight sends it scurrying back into its hole for six more weeks, during which time we get an extended winter.

Groundhog Day has spawned festivals all across America. One of these was even the setting for a recent Bill Murray comedy film. Of course, Murray's character had a truly unusual Groundhog Day experience—his day was repeated over and over and over. But what about those of us who pass through February 2 but once a year?

A Canadian named Reuben Hornstein decided to conduct an official study of weather records to verify the Groundhog Day legend. He found that in three years out of ten, Toronto was overcast at noon on February 2, meaning that no groundhog would have seen its shadow at the traditional hour. The theory then is that the groundhog would stay out and spring would soon arrive.

But was winter over any sooner on those cloudy Groundhog Days? Not even once.

Okay, but the furry meteorologist did see his shadow in seven out of ten years. What happened during those winters? Four out of ten were milder and shorter than usual, despite six weeks of extra napping time for the supposedly frightened weather idol.

During the remaining three sunny years out of ten, the legend bore out: Winter dragged on longer than usual. So, in Toronto, in the land that adopted the groundhog as a second-stringer for the hedgehog, the Groundhog Day tradition provided an accurate forecast for only three years out of ten.

Folklore about Other Animals

The groundhog story alone could be proof that before the development of sophisticated instruments, computers, and the local television weatherperson, farmers and sailors and everyone else were starved for clues that would help them predict the weather. Many people still accept the idea that animals are better about predicting nature; that is why weather sayings and folklore involving animals are so common. Along with this "wisdom" came a lot of superstition. Even today it is sometimes difficult to separate the facts from the folklore. After all, weather-predicting goats have made headlines even in today's high-tech world!

Comparing animals to weather instruments—barometers, thermometers, hygrometers, and weather vanes—can help us understand which beliefs are true. Because most species are vulnerable to environmental change, they react to even the slightest shift in air pressure, humidity, temperature, or wind direction.

This explains why some animals, such as horses and cows, become irritated when pressure falls; and why swallows fly lower when air pressure drops. A drop in air pressure indicates inclement weather.

Dogs use their sense of smell for hunting. When the air is moist and humid, as it is before a rain, the smell organs in dogs' noses become more sensitive.

Yet these animals are merely reacting to existing environmental conditions, not predicting the weather. Humans also experience physiological changes induced by changing weather conditions; but, for many reasons, we don't react as strongly as animals do. We spend much more time indoors than do most animals, and our survival does not depend so much on being hypersensitive to weather changes.

But what about the ability to make long-term forecasts that often is ascribed to animals? None of these capabilities is known to be useful for accurate long-range weather planning. For example, a squirrel's food supply will not tell what the weather will be like in three months. Another ancient long-term weather-prediction saw holds that a tough winter lies ahead if a bear or a horse is found to have a thick coat in early winter. We know now that a thick coat means simply that the early winter was cold and not that the rest of the winter will be harsh. As proof: December 1989 was the fourth-coldest December in history but was followed by the warmest January on record. Still, some of these ancient adages are fun, even today. The following are a few we've collected:

- When a dog rolls on his back, it soon will rain.
- When horses are restless and paw with their hoof, You'll soon hear the patter of rain on your roof.
- Watch the heifer's tail; When stretched aloft, 'twill rain or hail.
- Dogs hunt better before a rain.
- Swallows fly high, clear blue sky.

Swallows fly low, rain we shall know.
- Never sell your hen on a rainy day.
- It will be a cold, snowy winter if:
  Squirrels accumulate huge stores of nuts.
  Hair on bears and horses is thick early in season.

Folklore for Those with a "Bug" about the Weather

Insects are extremely sensitive to changes in their environments. All kinds of changes occur in the insect world as its countless billions of inhabitants respond to the slightest change in weather conditions. Well before you can sense subtle atmospheric changes, you may be able to notice changes in the behavior of nearby insects. Therefore, many of these reactions can be used to make accurate short-term weather forecasts.

The following are a few adages about insects and weather that are factually based. Some have been known and used for centuries yet stand up today as accurate measuring devices.

- When air pressure drops, insects become more active.
- In dry weather cicadas hum loudly, using their stiff, dry wings; but in high humidity their wings soften and can't vibrate.
- Bees cannot carry pollen easily when it is humid, and so they buzz furiously in frustration.
- Spider webs are sensitive to humidity, and after absorbing moisture from the air, the threads of the web break. So when a spider spins a web in dry weather, the web self-destructs as soon as the air becomes moist.
- Crickets chirp by sawing on their bodies with their back legs. The higher the temperature, the faster they saw.

While today we know them to be technically true, these explanations don't make interesting weather sayings. Long ago, however, people summarized these facts into short, easy-to-recite verses, as the following examples show:

- Ants are very busy, gnats will bite;
  Crickets are lively, spiders leave their nest;
  And flies gather in houses before a rain.
- Locusts sing when the air is hot and dry.
- When spider webs in air do fly,
  The spell will soon be very dry.

There is no scientific proof that insects can make long-range weather predictions. They simply are highly sensitive observers of current weather conditions. How about plants, though? Certainly their survival is closely tied to the weather. The ability to predict long-range weather would be a most desirable talent for a plant to own.

Plant Meteorologists

Plants interact with the weather in many ways to live and grow and are very sensitive to climate. Plants continually adjust to the changing weather and often respond visibly, especially to humidity. Many flowers close up as humidity rises, to protect their pollen from washing away. Many trees curl their leaves upward just before a storm. Seaweed and moss easily absorb moisture.

But the plant traits that are used to predict weather have actually occurred from weather that has already happened. Plants respond very slowly to weather and are of no use in making predictions, either short-term or long-term. For example, a thick onion skin is the result of a dry summer, not a sign of an upcoming harsh winter. So you can skip the following weather adages when you put together your weather forecasting kit:

- When the down of a dandelion contracts, it is the sign of rain.
- Sensitive plants contract their leaves at the coming of rain.
- Cottonwoods turn up their leaves before a rain.
- If onion skins are very thin,
  Then winter's mild when coming in.
  But if onion skins are thick and tough,

> ## Planting Lore
>
> Go plant the bean when the moon is light,
> And you will find that this is right;
> Plant the potatoes when the moon is dark,
> And to this line you always hark.
>
> But if you vary from this rule,
> You will find you are a fool;
> If you always follow this rule to the end,
> You will always have money to spend.

Then winter's long, cold, and rough.
- The daisy shuts its eye before rain.
- Seaweed dry, sunny sky;
  Seaweed wet, rain you'll get.

Noses, Bones, and Other Forecasting Tools

Like animals, insects, and plants, humans are dependent on interpreting changes in the environment for weather forecasts. Although we have sophisticated computers to pump out materials for forecasts, our senses can teach us to recognize potential weather changes. Through physics we can learn how weather modifies the physical behavior of some of Earth's elements. Knowing those effects, we know what to look for to help us predict weather by carefully observing our local environment. Here are some facts from physics about climatic changes that we can sense:

- Sound travels better in higher humidity. People in cities have long gauged the chance of rain by the clarity with which they can hear church bells ringing in the distance.
- When the air is colder, the squeaks of footsteps on snow are much louder.
- Water vapor has no odor, but approaching rain can be smelled. Plants ooze oils into the soil; and when air pressure drops, the soil releases some of these oils, producing the "smell of rain."
- Dropping air pressure and moisture in the air also amplifies the odors of flowers and manure enough that we can sense the difference.
- The optical illusion that faraway objects appear closer before a storm is caused by a difference in temperatures between levels of air. When upper air is warmer than normal, the light is refracted differently, distorting the view. This mirage is called the Hillenger Effect.
- Some people feel rain in their bones, especially if they suffer from arthritis or rheumatism. Numerous studies have found that a drop in barometric pressure brings discomfort to people who suffer joint diseases.

We now have scientific proof of the weather's effect on our senses; but long before the facts listed above were proven, people had created sayings that embodied the principles. Often using the form of short rhymes, people have turned raw scientific facts into accurate and easily remembered sayings such as the following:

- The minds of men do in the weather share,
  Dark or serene as the day is foul or fair.
- A coming storm your shooting corns presage,
  And aches will throb, your hollow tooth will rage.
- I know ladies by the score,
  Whose hair, like seaweed, sense the storm;
  Long, long before it starts to pour,
  Their locks assume a baneful form.
- If corns and bunions ache all day,
  You know that rain is on the way.
- Sound traveling far and wide,
  A stormy day will betide.
- Distant objects appear to be closer before rain.
- The squeak of the snow
  Will the temperature show.
- When ditches and ponds offend the nose,
  Look for rain and stormy blows.

- Flowers and manure piles smell stronger before it rains.
- Old Betty's nerves are on the rack,
  I think the rain is coming back.
- When folks complain more about pain,
  Expect the morrow to be filled with rain.

Linking Weather to Patterns

The rising and setting of the sun, the phases of the moon, the movements of the ocean tides—these all occur as regular patterns in nature. Weather, however, is not so predictable.

Still, over the centuries people have created hundreds of sayings that attempt to link the stars, the moon, or the first snow to weather events. Here are some examples:

- If it thunders on April Fool's Day,
  It will bring good crops of corn and hay.
- If it rains on St. Philip's and St. James', a fertile year may be expected.
- If it rains on Easter Sunday, it will rain for seven Sundays in a row.
- Easter in snow,
  Christmas in mud.
- St. Swithin's Day [July 15] if thou dost rain,
  For forty days it will remain.
  St. Swithin's Day if thou be fair,
  For forty days 'twill rain na mair.
- As many days old as is the moon on the day of the first snow, there will be that many snowfalls by crop planting time.
- If there is no snow in January, there will be more in March and April.
- If Candlemas [February 2] be fair and clear,
  There'll be two winters in the year.

Weather folklore also often states that nature seeks a balance (e.g., a hot summer indicates a cold winter and a dry spring means a rainy summer). But nature isn't so consistent when it comes to long-range forecasting. Overall, balancing may occur across the centuries, but you won't see it from season to season.

Cultural Diversity

Weather lore and weather sayings are a part of every culture. Here's a collection of weather sayings sampled from cultures around the world:

- English: High winds blow on high hills.
- Swedish: No one thinks of snow that fell last year.
- African: If you sow to the wind, you will reap the hurricane.
- Chinese: Let each sweep the snow from his own door; let him not be concerned about the frost on his neighbor's tiles.
- Japanese: The wind that comes in through a crack is cold.
- Filipino: Don't empty your water jar until the rain falls.
- Jewish: Rain chases you into the house; a quarrelsome wife chases you out.
- Greek: If the wind is not on your road, let it blow.
- Korean: Falling raindrops will wear through a stone.
- Hungarian: The wind will fell an oak but can not destroy the reed.
- Indian: Though the heavy rain is over, the dropping from the trees continues.
- Armenian: What the wind brings it also will take away.
- Albanian: Who has been almost drowned fears not the rain.
- French: When there is no wind, every man is a pilot.
- Italian: From snow, whether cooked or pounded, you get nothing but water.
- Turkish: However much snow falls, it does not endure the summer.
- Spanish: Wind and good luck are seldom lasting.
- Persian: The larger a man's roof, the more snow it collects.

—Courtesy of Rod Phillips,
Stormfax® Weather Services

Flood Folklore

Native Americans had a great deal of cultural diversity even among individual tribes and nations. Just in the one area of lore about floods, there were many tales.

The Utes called themselves the People of the Shining Mountains. Living in the Rocky Mountain area, they passed down a story about Pike's Peak remaining above the waters of a worldwide flooding of the Colorado River.

Other Native American tribes have great flood stories, too. A Chitimachan tale from southern Louisiana is like many of them. All the people except a man and a woman were destroyed by a huge flood. They were saved when they made a large eathern vessel and a dove brought them a single grain of sand from which a new land grew.

In a Pagago story, the boulders on the mountaintops are the rocky forms of people who asked the Great Spirit to save them from the pain of drowning in the great flood.

Cultural Craziness: The Stormfax® American Movie Weather Quiz

Even though cultures all over the world have had weather lore for centuries, all of them together have not cranked out more words, on any topic, than the information-overloaded modern-day U.S. of A. Here's a fun movie trivia quiz that will show you exactly how absorbed Americans can get in two of their favorite topics: movies and the weather.

1. In what 1952 movie does Gene Kelly give bad weather a starring role?
2. In the 1948 movie *Key Largo* with Humphrey Bogart and Lauren Bacall, what actor and what weather hold everyone hostage?
3. In what 1942 movie does Bing Crosby initially refuse to sing a song that later became his biggest-selling record? What is the song?
4. In the 1947 film *It's A Wonderful Life*, how does Jimmy Stewart know he's back home in Bedford Falls?
5. How do Dorothy and Toto get from Kansas to Oz in the 1939 classic *The Wizard of Oz*?
6. In the 1940 classic *The Grapes of Wrath*, with Henry Fonda, why are all the Oklahoma farmers moving to California?
7. What is the title of the 1956 movie that stars Katharine Hepburn as a woman romanced by con man Burt Lancaster?
8. Another movie couple, Clark Gable and Claudette Colbert, find their romance peaking in the rain during a bus trip in what Oscar-sweeping 1934 film?
9. Trapped in a cabin by a blizzard in the Yukon, Charlie Chaplin cooks one of his shoes for food in what 1925 silent comedy classic?
10. William Hurt and Kathleen Turner plot to murder her wealthy husband in what 1981 suspense drama set in sweltering Florida weather?
11. Dorothy Lamour and Jon Hall starred in a classic 1937 movie that many film critics say has never been equaled by modern special effects in depicting violent weather. What's the title of this island drama?
12. Silent comedy star Buster Keaton gets caught in a cyclone and turns his wind troubles into one sight gag after another near the end of what 1928 movie?
13. Off-season hotel caretaker Jack Nicholson goes berserk when deep snow cuts him and his family off from the outside world in what 1980 Stephen King thriller?
14. Aging camp director Uncle Lou, played by Alan Arkin, invites some of his favorite campers, now grown up, to return for a reunion before the camp closes for good. The old rules still apply: Everyone up at dawn, and boys and girls in separate cabins. Name this 1993 comedy-drama that takes place at Camp Tamakwa, named for an Indian weather god.

CHAPTER 2: WEATHER LORE

15. Audrey Hepburn and George Peppard star in this 1961 love story about a small-town girl who eventually finds happiness in Manhattan, as she and her cat are drenched by rain, to the tune of "Moon River." What's the name of this film, based on a Truman Capote story?
16. In the 1933 H. G. Wells fantasy *The Invisible Man*, Claude Rains is a mad scientist who wreaks havoc on a small British village until what weather-related event leads to his capture?
17. In the sentimental 1984 film *The Natural*, gifted baseball player Robert Redford finds there's something almost magical about a homemade bat fashioned from what?
18. Picture this: One day a giant electrical traffic warning sign starts sending personal messages to a wacky TV weatherman, suggesting how he can improve his life! The writer and star of this 1991 comedy worked on the screenplay for seven years. Name the actor and the movie.
19. In the 1931 movie *Frankenstein*, a mad scientist needs to give his creature (played by Boris Karloff) a jump start to get him going. What does he do?
20. In what 1985 comedy does another crazed scientist, played by Christopher Lloyd, need the help of lightning to zap his creation into action?
21. Lightning provides the inspiration to turn Gary Cooper from a drafted World War I pacifist into a hero in what 1941 film that won Cooper an Oscar?
22. When you live in the penthouse of an old New York City apartment complex, you worry about lightning storms. However, if you're Sigourney Weaver and there's lightning in your refrigerator, who ya gonna call?
23. In this 1933 classic, Skull Island would have been found sooner but is shrouded by a year-round fog bank. Name the film that also gave rise to the expression, "Ya big ape!"
24. This 1964 musical starred Rex Harrison and Audrey Hepburn, won eight Oscars and refined our speech with a tune about weather. What are the titles of the film and the song?
25. When it's pouring rain and you can hardly see to drive, you naturally pull into the nearest motel for the night. Right? Wrong! What infamous motel does Hitchcock's 1960 thriller *Psycho* teach us to avoid, even if it is storming?
26. At the age of twenty-five, Orson Welles made movie history with his first film. The fascinating story of a publisher's rise to power begins and ends with a childhood memory of sledding in the snow. Name this 1941 classic and name the sled.
27. When it comes to movie titles, the 1943 musical *Stormy Weather* tops the list for most meteorologists. The film also produced a title song that was a hit for what singer and star of the movie?
28. Humphrey Bogart and Katharine Hepburn battle the elements and each other as they travel the Congo River during World War I in what 1951 film? (Hint: Overnight rain floats their snagged boat and frees their feelings.)
29. What's the name of the shipboard-school vessel in the 1996 movie *White Squall,* starring Jeff Bridges as the teacher/skipper?
30. The famous London fog has trapped many a movie criminal. In what 1956 suspense film starring Van Johnson and Vera Miles do the fog and our hero trap the killer? Van's character has the advantage in fog—he's blind.
31. This 1955 musical stars Gene Kelly and Cyd Charisse in a story of three buddies who get together ten years after World War II to find they now have nothing else in common.
32. Paul Newman and Joanne Woodward made their first film together in 1958. Orson Welles plays Woodward's domineering father, who tries to keep Newman from marrying his daughter. Name this sultry Southern drama.
33. Right after his film *Halloween*, John Carpenter directed this 1980 thriller about a 100-year-old ghost ship and a killer fog that haunt a coastal California town. You get three guesses for the movie's title.
34. In the 1928 silent epic *The Wind*, a young girl battles a desert storm and a couple of bad guys. Who's the actress?

35. Actress Janet Gaynor makes her film debut in this silent-era disaster film from 1926 that still rivals today's special effects.
36. Speaking of special effects, this 1939 movie earned an Academy Award for its earthquake scenes. Name the film that brings together Tyrone Power, Myrna Loy, and heavy downpours.
37. A remake of the above film in 1955 adds color, Cinemascope, and stars Richard Burton sweating torrents over Lana Turner. What's the name of the movie?
38. In this 1993 Oscar®-winning, special-effects-packed movie, a hurricane is responsible for turning dinosaurs loose in an island amusement park. Sam Neill, Laura Dern and Jeff Goldblum play the pursued paleontologists in this Stephen Spielberg hit.
39. Pursuing tornadoes has Helen Hunt and Bill Paxton dodging flying cows and racing "weather bad guys" in this meteorologically scary movie written by Michael Crichton. What's the name of this 1996 film that glamorized storm chasing?
40. In what may be the first film about a weather man, Richard Widmark plays a former Navy weather pilot who takes time to think back on his life and loves (Linda Darnell and Veronica Lake) while flying through a storm. Name this 1949 movie.
41. Walt Disney's *Fantasia* made use of multichannel stereophonic sound and set a storm to music in 1940. The animation featured Zeus, the Greek god of the elements and his thunderbolt-maker Vulcan. Name the classic Beethoven musical expression of the spirit of nature.
42. While loosely related to weather in title only, name this pioneering 1937 Disney-animated feature film that taught us never to take apples from strangers.
43. TV weatherman Bill Murray travels to a western Pennsylvania hamlet and finds himself falling in love with Andie MacDowell, all the while reliving the same 24 hours over and over again. What's the title of this clever 1993 comedy-fantasy? For extra credit, spell the name of the tiny town.
44. Speaking of TV weatherpeople, Nicole Kidman won a Golden Globe Award for Best Actress in a Comedy for her 1995 role as a local TV weatherperson. She'll do anything to be on television, even plot to have her husband, played by Matt Dillon, murdered. Name this dark comedy which was inspired by true events.
45. John Travolta gets a jolt from a bolt of lightning that turns him into a chess expert, a language master, and gives him supernatural powers. Name this charming 1996 film reminiscent of Frank Capra movies.

—By Rod Phillips, Stormfax® Weather Services

Answers to the Stormfax® American Movie Weather Quiz

1. *Singin' in the Rain*
2. Edward G. Robinson and a hurricane
3. *Holiday Inn*; the song was "White Christmas"
4. It's snowing again.
5. In an unusually mild-mannered tornado
6. A drought has turned their land into a dust bowl.
7. *The Rainmaker*
8. *It Happened One Night*
9. *The Gold Rush*
10. *Body Heat*
11. *The Hurricane*
12. *Steamboat Bill, Jr.*
13. *The Shining*
14. *Indian Summer*
15. *Breakfast at Tiffany's*
16. He leaves his footprints in the snow.
17. Wood from a tree struck by lightning
18. Steve Martin in *L.A. Story*
19. He hoists the creature to the top of a tower, where a lightning strike provides the necessary jolt.
20. *Back to the Future*
21. *Sergeant York*
22. *Ghostbusters* (1984)
23. *King Kong*
24. *My Fair Lady*; "The Rain in Spain"
25. The Bates Motel
26. *Citizen Kane*; "Rosebud"
27. Lena Horne
28. *The African Queen*
29. *Albatross*
30. *23 Paces to Baker Street*
31. *It's Always Fair Weather*
32. *The Long, Hot Summer*
33. *The Fog*
34. Lillian Gish
35. *The Johnstown Flood*
36. *The Rains Came*
37. *The Rains of Ranchipur*
38. *Jurassic Park*
39. *Twister*
40. *Slattery's Hurricane*
41. "The Pastoral Symphony" (Beethoven's 6th)
42. *Snow White and the Seven Dwarfs*
43. *Groundhog Day*; Punxsutawney
44. *To Die For*
45. *Phenomenon*

Chapter 3
Violent Weather

In the realm of the skies of Earth, the supreme ruler is violent weather. Mankind never has been able to conquer it and likely never will. Violent storms are like airline crashes: They don't happen often, but when they do they make headlines and can command our full attention.

Violent storms are similar to plane crashes in other ways, too. They are triggered by a chain of events, none of which alone would produce violent weather but that together cascade into one spectacular weather event. The key ingredients in violent weather are always the same, though: temperature to spark the explosion, moisture to fuel the engine, and wind to stir things up.

It's amazing how these few key ingredients can combine in so many different ways to create such a wide variety of violent weather: blizzards, tornadoes, hurricanes, thunderstorms, and gales. All have the same basic components. Yet when they are mixed together at different temperatures, they produce vastly different results.

For this chapter we've collected three stories about three types of storms, each exceptional in itself. As you read the accounts, you'll see how factors combine in unusual ways to push things over the line that separates merely nasty weather from a truly violent storm. We'll give you one story each on hurricanes, tornadoes, and blizzards.

We also have included other stories about violent weather, scattered throughout the book. In later chapters, for example, you'll read about such phenomena as the unique conditions that combined to create the massive snowstorm on the East Coast in March 1993—dubbed by many "The Storm of the Century."

You also will find tips on how to predict, detect, and track violent storms so you can prepare for their arrival. And you'll learn what to do when and if they actually hit your hometown. We'll wrap up with a listing of some weather products that will help you employ the weather-tracking tips and lessons provided in this chapter. Let's start out by learning how a destructive monster hurricane can form and what happens to all in its path.

Hurricanes

If we likened violent storms to powerhouse football teams, hurricanes would be the Super Bowl champions of each season. During the regular season tornadoes might deliver some more spectacular games than the "big one"; but over the long haul, a hurricane will leave the most lasting impression and define the season. (Chapter 4, "Atmospheric Pressure," includes an article that gives a clear explanation of how a hurricane can become a monster.)

When most people think of hurricanes, they think of Florida, the Caribbean, or the Gulf Coast.

Certainly the hurricanes that occur in those places have been in the news enough that nearly everyone can recall a hurricane or two that struck one or more of those areas. Hugo or Andrew may come to mind.

Many people who live outside those areas wonder why anyone would live in "hurricane country" and probably are happy that they are "safe" from hurricanes. Yet the most populous region in the United States—and one that lies well north of normal hurricane tracks—can take a direct hit by a storm that should be only "tropical."

More than sixty years ago, the Chesapeake–Potomac Hurricane set a record as the most damaging storm to strike the Mid-Atlantic region of the United States. The dollar-damage reports of $40 million may seem unimpressive today, but remember that back then a new car cost only a few hundred dollars and that few houses cost as much as a used car does today. Imagine what would happen to the damage record set by Hurricane Andrew if a storm followed the same track and packed the same power today as did the hurricane in 1933.

Trees Swayed by Hurricane-Force Winds

Hurricane Classifications

Category 1: Winds between 74 and 95 mph and/or a storm surge of 4 to 5 feet above normal
Category 2: Winds between 96 and 110 mph and/or a storm surge of 6 to 8 feet above normal
Category 3: Winds between 111 and 130 mph and/or a storm surge of 9 to 12 feet above normal
Category 4: Winds between 131 and 150 mph and/or a storm surge of 13 to 18 feet above normal
Category 5: Winds greater than 150 mph and/or a storm surge greater than 18 feet above normal

History of Hurricane Tracking

The first hurricane warning in the United States was flashed in 1873, when the Signal Corps warned against a storm approaching the coast between Cape May, New Jersey, and New London, Connecticut. Today, classifying and naming a storm launches a far more sophisticated warning system that activates long-distance communications lines and tested preparedness plans.

The day is long gone that a hurricane could develop to maturity far out at sea and then attack the coastlines without warning. Earth-orbiting satellites operated by the NOAA keep the earth's atmosphere under virtually continuous surveillance, night and day. Long before a storm has evolved even to the point of ruffling the easterly wave, scientists at NOAA's National Hurricane Center in Miami, Florida have begun to watch the disturbance.

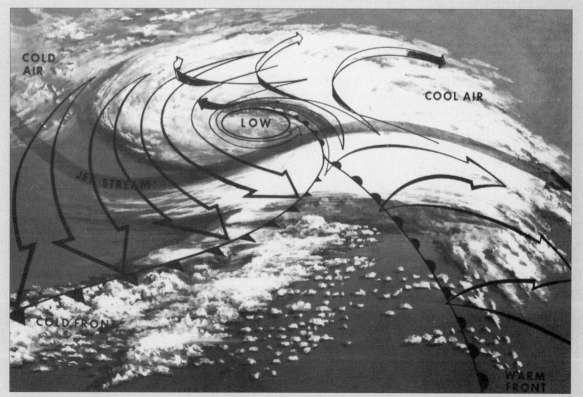

Modern tracking systems include detailed satellite photographs.

Chesapeake—Potomac

The hurricane of August 23, 1933, is the only one to make a direct hit on the Mid-Atlantic region this century. Though it was only a weakening "moderate" hurricane at the time of landfall, an unusual pressure pattern made the storm the most damaging in the history of the area. It killed 47 people, caused $40 million damage (in Depression dollars) and devastated coastal areas from Virginia to New Jersey. The hurricane had its greatest impact, however, in the Chesapeake Bay–Potomac River basin area.

The Mid-Atlantic area does not usually suffer direct hits from land-falling hurricanes. To make a direct hit, a hurricane must approach from the southeast, an unusual trajectory for a tropical cyclone traveling in the zone of prevailing westerlies, where it would tend to recurve to the north and northeast. This hurricane, however, moved in a general northwesterly direction for its entire lifetime.

The Chesapeake–Potomac hurricane developed near 18° N, 50° W on August 17, 1933, in the midst of the most active Atlantic hurricane season on record; a total of 21 tropical storms and hurricanes developed that year.

A Far-Fetched Pressure Pattern

On August 19, 1933, a large high-pressure area moved southeastward from Quebec to Maine. Further south a frontal system had pushed off the North Carolina coast and stalled just offshore. A strong gradient between high pressure to the north and lower pressure along a stalled frontal boundary produced a long easterly fetch of wind and wave over the open Atlantic from 55° W to the Mid-Atlantic coast. As a result tides rose along the entire coastal region on August 20. In the meantime, ship reports showed that the hurricane, now centered near 27° N, 60° W was intensifying, with an estimated central pressure of 27.76 inches (940 mb). That would support winds of at least 135 mph. Thus at this time the storm was a "very strong" hurricane.

The low-level easterly flow and associated overrunning conditions continued unabated into August 21, resulting in torrential rains throughout the Mid-Atlantic region. Rainfall totals for the 24 hours ending at 8 A.M. August 22 broke all previous records in Norfolk, Virginia, and in Atlantic City, New Jersey, where 6.54 inches and 8.30 inches fell, respectively. More than 4 inches were logged in Washington, D.C., Philadelphia, and Baltimore.

At noon August 21, as weather conditions deteriorated in the Mid-Atlantic region, the hurricane was sweeping 100 miles southwest of Hamilton, Bermuda. The storm was moving steadily northwest at 20 mph, and winds in the port city were gusting to 80 mph. Such strong winds 100 miles north of the storm center were an indication of the strength of the easterly flow between the high pressure to the north and the advancing hurricane to the southeast. Despite the hurricane force gusts, hundreds of curious onlookers crowded the beaches to watch the heavy surf wreak havoc on fishing piers and nearby cottages.

As the tropical cyclone reached full fury in Hamilton, Bermuda, the U.S. Weather Bureau issued northeast storm warnings for coastal areas from Cape Hatteras, North Carolina, to Boston, Massachusetts, effective 10 A.M. August 21. This was only a precautionary move just in case the storm should brush the Mid-Atlantic coast during its expected recurvature to the northeast. Unfortunately, however, the mound of high pressure over Maine blocked a northward turn and recurvature of the storm. Moreover, the remnants of the low-pressure frontal system provided a path of least resistance for the hurricane. So the storm turned more to the west and headed straight for Cape Hatteras, North Carolina.

Weather conditions worsened all along the Mid-Atlantic coast during the morning of August 22. By mid-morning gale-force winds were raking the outer banks of North Carolina, and tides were running up to 4 feet above normal northward to New Jersey, inundating low-lying areas.

Longtime residents sensed this would be no ordinary storm and began to make emergency preparations.

Devastation began along the North Carolina coast by sunset. A little later ship reports indicated that the storm had weakened considerably and now had an estimated central pressure at 28.35 inches (960 mb).

By 9 P.M. the hurricane was centered 100 miles southeast of Cape Hatteras, churning northwest at 17 mph. The eye sideswiped Cape Hatteras shortly after 3 A.M. August 23. Winds peaked at 80 mph while the barometer read 28.67 inches (971.3 mb). At the same time the SS *Hanna* encountered the full brunt of the storm near 35.5° N, 75.0° W, where the ship's barometer recorded 28.54 inches (966.6 mb). Based on this report, the central pressure was very close to 28.50 inches (954.4 mb), so the storm was still weakening. It would continue to weaken, but because of its interaction with the high-pressure pattern to the north, it would cause unprecedented damage in the Mid-Atlantic region.

The hurricane made landfall near Nags Head, North Carolina, shortly after 4 A.M. The dangerous semicircle of the storm—where the hurricane's forward speed was added to the counterclockwise hurricane winds—swept over the beach highway between Nags Head and Kitty Hawk, washing it out for three miles. At the height of the storm surge, the waters of the Atlantic Ocean joined those of Albemarle Sound. Several cottages, one of them sheltering a family of six, were carried into the sound. Based on the sparse reports from the area, the winds shifted from the northeast to the south-southwest shortly after 6 A.M., indicating the storm was still moving northwestward, into southeastern Virginia.

The Chesapeake–Potomac hurricane was a major meteorological event in Norfolk, Virginia. A full-blown hurricane had not passed directly over the city since the Norfolk–Long Island Hurricane of September 3, 1821. At 6:35 A.M. sustained winds peaked at 57 mph at the Weather Bureau office downtown, with gusts near 70 mph. The Naval Air Station reported gusts to 88 mph. The barometer plunged to 28.68 inches (971.6 mb) at 9:20 A.M., when the eye of the storm was right over the city. This was a record low reading for tropical storms in the area.

When the leaden skies finally cleared, the waters of the Elizabeth River, piled up by 18 hours of gale-force winds, were flowing over the docks at Main Street and up to 1.5 miles inland, inundating the entire business district. Water depths ranged from 1 foot to nearly 5 feet in a few low places. Swimming was the sport of the day on Granby Street. The official tide gauge at Sewells Point recorded a high mark of 9.79 feet above mean low water, the highest ever recorded there. Tides were even higher—perhaps as much as 12 feet—in some of the narrower estuaries in the Norfolk area.

Retired Hurricane Names

Most hurricane names are recycled, but some cause so much damage that the name is retired. Here's a listing of retired hurricane names over the years:

1954: Carol, Edna, Hazel	1966: Inez	1983: Alicia
1955: Connie, Diane, Ione, Janet	1967: Beulah	1985: Elena, Gloria
1957: Audrey	1969: Camille	1988: Gilbert, Joan
1959: Gracie	1970: Celia	1989: Hugo
1960: Donna	1972: Agnes	1990: Diana, Klaus
1961: Carla, Hattie	1974: Carmen	1991: Bob
1963: Flora	1975: Eloise	1992: Andrew
1964: Cleo, Dora, Hilda	1977: Anita	1994: Gordon
1965: Betsy	1979: David, Frederic	1995: Luis, Marilyn, Opal, Roxanne
	1980: Allen	

Chesapeake Bay Tidal Bore

As the eye of the storm moved north of Norfolk, a Chesapeake Bay tidal bore (a high-breaking wave that moves rapidly up an estuary) began to develop. The convergence of bay and ocean waters in the vicinity of Norfolk produced a huge mound of water that moved up the bay, increasing in height, especially after the wind shifted to the south. Since the hurricane itself moved up the entire length of Chesapeake Bay at a speed only slightly greater than that of the propagation of such shallow water waves, it continued to feed energy into this storm-surge phenomenon, resulting in its great amplification.

As the hurricane swept east of Richmond, Virginia, in the early afternoon hours of August 23, the tidal bore smashed into the Northern Neck of Virginia. Driven by hurricane-force southeasterly winds, the Potomac River went on a rampage, pouring into the village of Colonial Beach. The amusement park was swept away, and hotels along the riverfront were inundated to a depth of 4 feet. The hurricane then turned its full fury upon the nation's capital.

The arrival of the hurricane in Washington, D.C., was heralded by 50 mph winds and torrents of rain. Rainfall totaled a record 6.18 inches for the 24-hour period ending at 7 P.M. August 23. Over 7 inches fell in neighboring Fairfax County to the west, resulting in severe crop damage.

The Chesapeake Bay–Potomac River tidal bore swept into the Washington area coincident with the passage of the storm shortly after 7 P.M. The barometer at the Washington-Hoover (now National) airport fell to 28.94 inches (980 mb). In Old Town, Alexandria, Potomac River waters rose to 12 feet above normal, the highest level ever recorded. The Alexandria Torpedo Factory and the Ford Motor Company were flooded to a depth of 6 feet at high tide.

The Washington–Richmond highway was submerged under 10 feet of water at the height of the storm. Further north, at Bolling Air Force Base, the hangars were inundated by 5 feet of river water. Meanwhile, the ever-driving storm surge shoved its way up the Anacostia River and swept a Washington–Philadelphia express train off the tracks as it was crossing the river. Ten people died.

The swollen Anacostia River surged into the surrounding Maryland suburbs and inundated every thoroughfare in Bladensburg. Four people were drowned on the Baltimore–Washington highway when their cars were swamped by the wild waters of the Little Patuxent River. All told, the storm killed 18 people in the Washington area. The 1933 hurricane was the worst storm to visit the area since the tropical storm of 1896.

Hurricane-induced waves wreak havoc on the coastline.

CHAPTER 3: VIOLENT WEATHER

Unprecedented Coastal Devastation

As we have seen, the fetch of wind and wave associated with the 1933 hurricane was of extreme length and duration. The focal point of its four-day assault was the coastal area of the Delmarva (Delaware, Maryland, Virginia) Peninsula and New Jersey. The result was such that, as the storm moved inland over Norfolk on the morning of August 23, the pressure gradient between the hurricane and the huge area of high pressure then over Nova Scotia caused the radii of maximum winds in the storm to greatly expand out to the northeast. As a result, nearly every coastal station from Virginia Beach, Virginia, northward through New Jersey reported hurricane-force winds. The high winds, combined with the massive storm surge built up from the time the storm had passed Bermuda two days earlier, devastated this 200-mile stretch of coastline unlike any storm in the history of the area before or since.

Virginia Beach was hit hard. Winds at Cape Henry reached a maximum value of 66 M.P.H. with gusts to 82 M.P.H. National Guardsmen waded in waters up to their armpits to rescue stranded residents along Atlantic Avenue.

The hurricane then churned further into Virginia, where its "dangerous" semicircle of winds swept across the upper reaches of the tidewater area. Newport News was swamped by tides ranging from 5 to 8 feet above any previously recorded; Gloucester Point was treated to a lashing unlike any in its history as a 10-foot storm tide swept away the town post office and drug store. All in all, damages in the tidewater area of Virginia exceeded five million, very depressed 1933 dollars. Fifteen people were killed, most of them by the storm surge.

The storm surge also breached the narrow barrier island of Ocean City, Maryland, forming the Sinepatuxent Inlet, which now links Sinepatuxent Bay with the Atlantic Ocean. Salisbury, Maryland, some 30 miles inland, became a haven for Ocean City evacuees. However, not even Salisbury escaped the effects of the hurricane and the relentless storm surge. Southeast winds gusting to 65 M.P.H. drove the waters of the Wicomico River into the streets of the city. In some areas the water was chest high. In the surrounding countryside the Elk, Choptank and Pocomoke rivers surged out of their banks, inundating farms and destroying crops worth millions of dollars.

The state of Delaware also fared badly. Southeast winds of hurricane force created 35-foot breakers that slammed into the Rehoboth Beach boardwalk, breaking it into splinters. At high tide the Henelopen Hotel was flooded to a depth of several feet. Power was knocked out to most residents for a week. Even the northern portion of the state did not escape. In Wilmington the waters of the Delaware River did an about-face at high tide and backed up into the waterfront business district. Water depths ranged up to 6 feet in a few low-lying areas.

The tidal flooding also pushed up the Delaware River into neighboring Philadelphia. At one point 10 square miles of southwest Philadelphia were swamped by the storm surge. The hardest hit area was Stonehouse Village, where water was running 5 feet deep in the streets. The Philadelphia airport was submerged at high tide.

The 1933 hurricane was the worst natural disaster in the history of New Jersey until that time. Gale-force easterly winds had continued unabated since August 20th. When the storm swept inland over Norfolk, 200 miles to the south, the wind peaked at 88 mph at Wildwood, New Jersey. Hurricane-force winds were also reported in Atlantic City and Cape May.

Damages were extreme throughout southern New Jersey. In Wildwood, the morning high tide submerged the main thoroughfare under 6 feet of water. Just before 10 A.M., August 23, the waters of the Atlantic Ocean breached the narrow island comprising Wildwood and merged with those of the Grassy Sound, threatening to inundate everything and everybody.

Sixty percent of Atlantic City was flooded by the morning high tide, and vacationers waded in waters up to their necks to escape the storm surge. In Ocean City, New Jersey, the waters of Great

Egg Harbor and the Atlantic Ocean threatened to eliminate that island as well. The huge hurricane waves also swept away the fishing piers at Cape May and Avalon.

The northern portion of New Jersey and the metropolitan New York City area escaped the brunt of the Chesapeake–Potomac hurricane. Winds in the area were in the 45 M.P.H. range with a few reports of 60 mph gusts on Long Island. Nevertheless the massive storm tide affected those areas, as well.

"Astonishing" wave heights were reported all along the south shore of Long Island, and tides ran 4 feet above normal at The Battery in New York City. Tidal flooding was also reported along the East River in Manhattan and the Bronx.

The pounding surf from this hurricane even affected Nantucket Island and the southern shores of Rhode Island and Massachusetts, where there were reports of drownings in the strong undertow on August 23.

By late evening on August 23, the 1933 hurricane was becoming a frightening memory for Mid-Atlantic residents. At midnight, August 24, the storm was located in central Pennsylvania, moving steadily northward and still weakening.

Overall, the Chesapeake-Potomac hurricane took 47 lives and caused $40 million damage. Figuring adjustments for the depressed dollars of 1933 and the massive increase in coastal development since then, the Chesapeake-Potomac hurricane would have been a several-billion-dollar storm if it struck today.

—By Hugh D. Cobb III. Reprinted with permission from *Weatherwise*.

A very active hurricane season

CHAPTER 3: VIOLENT WEATHER

Recent Billion-Dollar Weather Disasters

Ever wonder about the highest losses ever caused by weather disasters? Perhaps you would name a famous hurricane as a candidate for the grand-prize winner. But the winner—and two of the top four—were not violent storms. Here's a list of the U.S. weather disasters over the past two decades that cost more than $1 billion, sent to us by Rod Phillips, chief meteorologist for the *Stormfax® Weather Almanac*:

- Flooding, December 1996–January 1997: California, Oregon, Washington, Idaho, and Nevada, $2.8 billion and 30 deaths.
- Hurricane Fran, September 1996: Virginia and North Carolina, $5.0 billion damage, 36 deaths.
- Southern Plains severe drought, Fall 1995–Summer 1996: Agricultural regions of Texas and Oklahoma were most severely affected, with $4.0+ billion damage, no deaths.
- Blizzard and Flooding, January 1996: A snowstorm over Appalachians, Mid-Atlantic region, and Northeast was followed by severe flooding in the same area from rain and snowmelt, with $3.0+ billion damage, 187 deaths.
- Hurricane Opal, October 1995: Florida panhandle, Alabama, western Georgia, eastern Tennessee, and the western Carolinas saw $3.0+ billion damage, 27 deaths.
- Hurricane Marilyn, September 1995: The U.S. Virgin Islands were devastated, with $2.1+ billion damage, 13 deaths.
- Texas/Louisiana/Mississippi flooding, May 1995: Torrential rain across the Dallas, Texas, area, southeast Louisiana (New Orleans was hardest hit), and southern Mississippi, with $5.0+ billion damage, 32 deaths.
- California flooding, January–March 1995: Frequent winter storms caused flooding across much of California, with $3.0+ billion damage, 27 deaths.
- Texas flooding, October 1994: Torrential rain and thunderstorms caused flooding across southeast Texas, with $1.0+ billion damage, 19 deaths.
- Tropical Storm Alberto, July 1994: Remnants of slow-moving Alberto brought torrential 10–25 inch rains, widespread flooding in Georgia, Alabama, and the Florida panhandle, with $1.0 billion damage, 32 deaths.
- Southeast ice storm, February 1994: Intense ice storm with damage in Texas, Oklahoma, Arkansas, Louisiana, Mississippi, Alabama, Tennessee, Georgia, North and South Carolinas, and Virginia, with $3.0 billion damage, 9 deaths.
- Southern California wildfires, Fall 1993: Wind-swept fires, resulting in $1.0 billion damage, 4 deaths.
- Midwest flooding, Summer 1993: Central United States, with $15–$20 billion damage, 48 deaths.
- Drought/heat wave, Summer 1993: Southeastern United States, with $1.0 billion damage, undetermined deaths.
- Storm/blizzard, March 1993: Eastern United States, with $3.0–$6.0 billion damage, 270 deaths.
- Nor'easter, December 1992: Slow-moving storm battered U.S. coast, New England hardest hit, with $1.0–$2.0 billion damage, 19 deaths.
- Hurricane Iniki, August 1992: Hawaiian island of Kauai, with $1.8 billion damage, 7 deaths.
- Hurricane Andrew, August 1992: Florida and Louisiana, with $27 billion damage, 58 deaths.
- Hurricane Bob, August 1991: Coastal North Carolina, Long Island, and New England, with $1.5 billion damage, 18 deaths.
- Hurricane Hugo, September 1989: North and South Carolinas, with $7.1 billion damage, 57 deaths.
- Drought/heat wave, Summer 1988: Central and Eastern United States, with $40 billion damage, 5,000 to 10,000 deaths.

- Hurricane Juan, October–November 1985: Louisiana and Southeast United States, with $1.5 billion damage, 63 deaths.
- Hurricane Elena, August–September 1985: Florida to Louisiana, with $1.3 billion damage, 4 deaths.
- Florida freeze, December 1983: Severe freeze in central and northern Florida, causing $2 billion in crop damage, no deaths.
- Hurricane Alicia, August 1983: Texas, with $3 billion damage, 21 deaths.
- Drought/heat wave, June–September 1980: Central and Eastern United States, with $20 billion damage, 1300 deaths.

One factor in calculating these damage estimates has been left out: how much these disasters cost in terms of missed work, tourist trade, and business opportunities; plus hotel, travel, and higher food costs for displaced victims. Recent discussions on weather-related computer bulletin boards indicate that weather experts are trying to work out guidelines to include those figures in damage estimates.

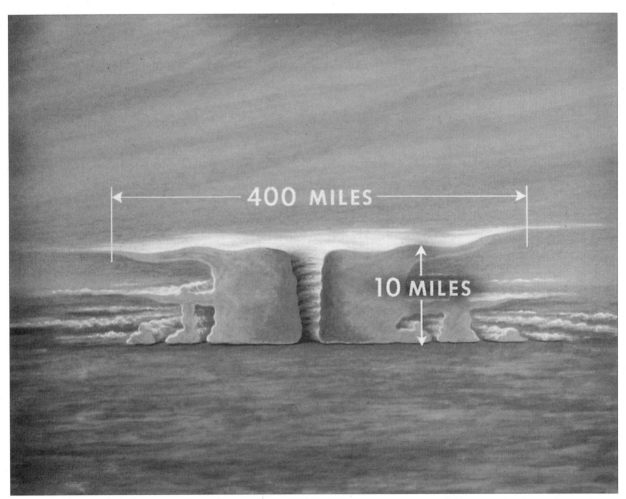

Hurricanes can span 400 miles.

CHAPTER 3: VIOLENT WEATHER

Storm Chasing

In the United States, the Indianapolis 500 motor race draws the largest crowd of any sporting event of the year. More than a half-million zealous racing fans migrate to the heartland of America to see this thrilling spectacle in person.

Imagine that instead of being scheduled for the Sunday before every Memorial Day, the Indy 500 were held randomly and that the location and site were a complete surprise to fans until the day the cars were ready to roar onto the track. Certainly such crazy circumstances would create a highly resourceful set of fans who would learn all they possibly could about the car owners and drivers so they could track their whereabouts. Inevitably, diligent fans would be rewarded when all their clues pointed toward a buildup in Indianapolis that indicated an Indy 500 was approaching. Those fans lucky enough to have figured out the clues would jump in their own cars and race to the track, hoping to beat the clock so they could see, feel, and hear the spectacle. And only a few of the most diligent would be rewarded.

That's how it is with tornado watchers. They're the zealous fans who resort to every weather-observation trick available so they can be among the few on hand when the powerful engine of a tornado roars into life. Few people can understand what motivates storm chasers, but those who've experienced chasing a tornado would never give up their "sport." (For an inside look at the adventures of one dedicated storm chaser, see p. 73.)

Here's a personal weather account from one of this book's authors, Ron Wagner, that he experienced while flying for the U.S. Air Force in the Presidential Wing at Andrews Air Force Base near Washington, D.C.

A standby flight crew is always available at Andrews AFB for emergency VIP travel. In June, 1979, I was the standby crew when I heard on the news that a devastating tornado had struck Wichita Falls, Texas, and Texas Senator John Tower was at the White House with President Carter to discuss emergency plans.

Assuming that the Senator would soon want to inspect the damage, I immediately headed for the flight line. A plane had already been towed to the VIP spot, and Senator Tower showed up a few minutes later with his personal secretary.

The line of storms that had spawned the tornado was so severe that to get to Wichita Falls, which is on the northern Texas border, I had to take the Senator out over the Gulf of Mexico—about 100 miles south of Houston and then turn north and fly clear across Texas.

Soon after passing Dallas, I received the most unorthodox call from a traffic controller that I've ever heard. Instead of normal, terse controller-speak, he sounded as if he were breaking bad news to a family member. "Pacer Zero One," he said, "I'd normally hand you off to approach control about now, but there's nobody out there for you to talk to. We've been calling them on land lines and can't get through. Since we knew you had Senator Tower on board, we've tried everything to get through and we finally found a local HAM radio operator who has spoken with some folks in Wichita Falls. They say the runway is clear of debris, you're the only traffic in the area and you can just go on in and land . . . but there won't be anyone on the radio . . . you're on your own. Good luck and give our best to the Senator."

We descended out of radio contact with Dallas and into the silence of Wichita Falls. There were no functioning navigation aids, but the Senator's secretary had once lived near the base and knew the lay of the land. She came into the cockpit to identify familiar landmarks as we sneaked through the dusty Texas sky. Approaching the edge of the city, she gasped in horror as we passed blocks and blocks of flattened houses—including ones on her sister's street—but she steeled herself to the task and actually guided us to the runway by referencing roads she recognized.

After landing, the Senator and his secretary were whisked away in a helicopter. Inside base operations, all the clocks were frozen at one unforgettable moment in time. With darkness closing in on a powerless Air Force base, they quickly refueled my plane and I escaped the scene of the worst destruction I have ever personally witnessed.

How Tornadoes Form and How to Stay Safe When They Do

Their time on Earth is short, and their destructive paths are rather small. Yet, when one of these short-lived, local storms marches through populated areas, it leaves a path of almost total destruction. In seconds, a tornado can reduce a thriving street to rubble.

It is the mission of NOAA, the U.S. Commerce Department's National Oceanic and Atmospheric Administration, to help mitigate the threat to life and property from natural hazards. The National Weather Service, a major element of NOAA, provides the Nation's first line of defense against the awesome destructive force of the tornado. Through its tornado and severe thunderstorm watches and warnings, the National Weather Service gives persons in threatened areas time to find shelter. Further, the National Weather Service, in cooperation with the Federal Emergency Management Agency (FEMA), educates community officials and the public through

Various Tornado Shapes

CHAPTER 3: VIOLENT WEATHER

its disaster preparedness program, on what to do when severe storms threaten.

This discussion is designed to increase your understanding and awareness of the tornado hazard, explain the National Weather Service Watch and Warning program, and provide you lifesaving safety precautions.

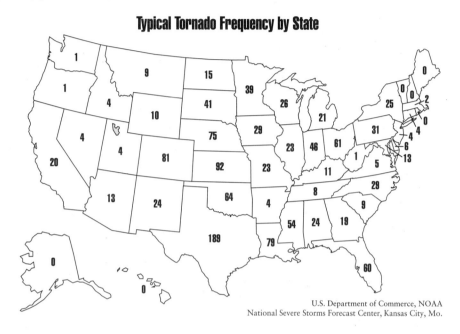

U.S. Department of Commerce, NOAA
National Severe Storms Forecast Center, Kansas City, Mo.

Tornado Characteristics

By definition, a tornado is a violently rotating column of air in contact with the ground. The air column may be seen when it contains condensation in form of a cloud or when it contains surface dust and debris. Often its appearance is a result of both. When a tornado touches the ground there usually is a swirl of dust and debris even when the visible cloud portion is missing or fails to reach all the way to the ground. When the column of air is aloft and does not produce damage, the visible portion is properly called a funnel cloud. A waterspout is a tornado in contact with a water surface.

Tornadoes vary greatly in size, intensity and appearance. Most (62 percent) of the tornadoes that occur each year fall into the weak category. Wind speeds are in the range of 100 mph or less. Weak tornadoes account for less than 3 percent of all tornado deaths. About one out every three tornadoes is classified as *strong*. Wind speeds reach about 200 mph with an average path length of 9 miles and a width of 200 yards. Almost 20 percent of all tornado deaths occur each year from this type of storm. Nearly 70 percent of all tornado fatalities result from *violent* tornadoes. Although very rare (only about 2 percent are violent), these extreme tornadoes can last for hours. Average path lengths and widths are 26 miles and 425 yards, respectively. The largest of these may exceed a mile or more in width, with wind speeds approaching 300 mph.

The color of a tornado is determined by a number of factors such as the amount and direction of sunlight and the type of debris being picked up at the surface. Not only does the shape of a tornado vary from one storm to another, but an individual tornado usually changes its shape frequently during its life cycle. During the late stages of a tornado's life, it is not unusual for the tornado to become highly tilted and shrink in size. This reduction in size does not mean that it is less intense. It is still very dangerous!

Tornado Rating System

If you saw the movie, *Twister*, you already know that tornadoes are rated for their destructive ability. In the movie, searchers were looking for an encounter with the dreaded "F5" level tornado, a rare, but incredibly dangerous storm that packs the most destructive power of any storm. Here's the official listing of the Fujita Scale that is the basis of the "F" ratings you hear about:

F0 (40–72 mph)
Damage is light and may include damage to tree branches, chimneys, and billboards. Shallow–rooted trees may be pushed over.

F1 (73–112 mph)
Damage is moderate. Mobile homes may be pushed off foundations and moving autos pushed off the road.

F2 (113–157 mph)
Damage is considerable. Roofs can be torn off houses, mobile homes demolished, and large trees uprooted.

F3 (158–206 mph)
Damage is severe. Even well–constructed homes may be torn apart, trees uprooted, and cars lifted off the ground.

F4 (207–260 mph)
Damage is devastating. Houses can be leveled and cars thrown great distances; objects become deadly missiles.

F5 (261–318 mph)
Damage is incredible. Structures are sucked off foundations and literally carried away. Cars become missiles. Less than 2 percent of all tornadoes generate this level of power.

F6 (318+ mph)
No recorded tornado has reached "F6" level.

With exceptionally large tornadoes, the classic "funnel shape" may be absent. The tornado may appear to be a large, turbulent cloud near the ground. It may even be mistaken for a large rain shaft or even a non-weather event such as a fire.

Sometimes a series of two or more tornadoes is associated with a parent thunderstorm. As the parent thunderstorm moves along, tornadoes may form, travel along in contact with the ground and dissipate or lift, followed shortly by other tornado touchdowns, and so on. Tornadoes can also be made up of a number of smaller but intense vortices that rotate about a common center.

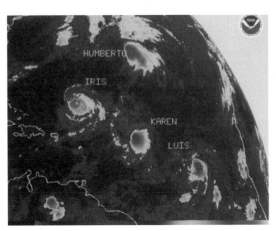

With this type, the most intense damage is concentrated along the paths of the small vortices.

While an individual tornado usually destroys a relatively small area, major tornado outbreaks may cause widespread damage over an extensive area. During the afternoon and evening of April 3 and the early morning of April 4, 1974, a "super outbreak" of 148 tornadoes across 13 states killed more than 300 people, injured more than 6,000 and caused $600 million in damage. On March 18, 1925, the Tri-State tornado traveled some

CHAPTER 3: VIOLENT WEATHER

219 miles across Missouri, Illinois, and Indiana. It lasted for 3½ hours and killed 689 people.

Except for weak tornadoes and waterspouts in coastal areas, tornadoes usually develop from strong or severe thunderstorms. Most significant tornadoes have their origin within the right-rear quadrant of the thunderstorm where a circulation develops at heights between 15,000 and 30,000 feet. A tornado or funnel cloud is observed when this circulation develops further downward toward the surface. Tornado development can also occur along the leading edge of a single thunderstorm or line of thunderstorms. While dangerous, such tornadoes are usually weak and short-lived.

Tornado Destruction

Every tornado is a potential killer and many are capable of great destruction. Tornadoes can topple buildings, roll mobile homes, uproot trees, hurl people and animals through the air for hundreds of yards and fill the air with lethal, windborne debris. Sticks, glass, roofing material, lawn furniture all become deadly missiles when driven by a tornado's winds. In 1975, a Mississippi tornado carried a home freezer for more than a mile. Tornadoes do their destructive work through the combined action of their strong rotary winds and the impact of windborne debris. In the most simple case, the force of the tornado's wind pushes the windward wall of a building inward. The roof is lifted up and the other walls fall outward. Until recently, this damage pattern led to the incorrect belief that the structure had exploded as a result of the atmospheric pressure drop associated with the tornado.

The Destructive Path of a Tornado

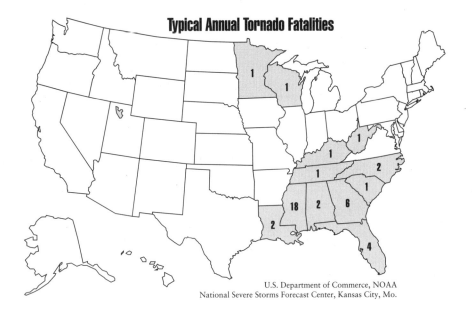

U.S. Department of Commerce, NOAA
National Severe Storms Forecast Center, Kansas City, Mo.

Mobile homes are particularly vulnerable to strong winds and windborne debris. Because they have relatively large surface-area-to-weight ratios, they are easily overturned by high winds. Their thin walls make them extremely vulnerable to windblown debris. Even if tied down, they should be evacuated for more substantial shelter. Mobile home parks should have storm shelters for their residents if located in areas where strong thunderstorms or tornadoes occur.

—Reprinted from *A Lesson on Tornadoes and Tornado Safety* by the National Oceanic and Atmospheric Administration.

Tornado Facts That Can Save Your Life

Tornadoes travel at an average speed of 30 mph, but speeds ranging from stationary to 70 mph have been reported. While most tornadoes move from the southwest to the northeast, their direction of travel can be erratic and may change suddenly.

In populated areas, it is very dangerous to attempt to flee to safety in an automobile. Over half the deaths in the Wichita Falls Tornado of 1979 were attributed to people trying to escape in motor vehicles. While chances of avoiding a tornado by driving away in a vehicle may be better in open country, it is still best in most cases to seek or remain in a sturdy shelter such as a house or building. Even a ditch or ravine offers better protection than a vehicle if more substantial shelter is not available.

- **While hail may or may not precede a tornado, the portion of a thunderstorm adjacent to large hail is often the area where strong-to-violent tornadoes are most likely to occur.**

Once large hail begins to fall, it is best to assume that a tornado may be nearby and seek appropriate shelter. Once the hail has stopped, remain in a protected area until the thunderstorm has moved away. This will usually be fifteen to thirty minutes after the hail ceases.

CHAPTER 3: VIOLENT WEATHER

- **The tornado's atmospheric-pressure drop plays, at most, a minor role in the damage process.**

    Most structures have sufficient venting to allow for the sudden drop in atmospheric pressure. Opening a window, once thought to be a way to minimize damage by allowing inside and outside atmospheric pressures to equalize, is not recommended. In fact, if a tornado gets close enough to a structure for the pressure drop to be experienced, the strong tornado winds probably already will have caused the most significant damage. Furthermore, opening the wrong window actually can increase damage.

- **While most tornado damage is caused by the violent winds, most tornado injuries and deaths result from flying debris.**

    Small rooms, such as closets or bathrooms, in the center of a home or building offer the greatest protection from flying objects. Such rooms are also less likely to experience roof collapse. Always stay away from windows and exterior doors.

- **Tornado wind speeds increase with height within the tornado.**

    Storm cellars or well-constructed basements offer the greatest protection from tornadoes. If neither is available, the lowest floor of any substantial structure offers the best alternative. In high-rise buildings it may not be practical for everyone to reach the lower floors, but the occupants should move as far down as possible and take shelter in interior small rooms or stairwells.

- **Tornado winds may produce a loud roar similar to that of a train or airplane.**

    At night or during heavy rain, the only clue to a tornado's presence may be its roar. Thunderstorms can also produce violent, straight-line winds that produce a similar sound. If any unusual roar is heard during threatening weather, it is best to take cover immediately.

- **Although most tornadoes occur during the midafternoon or early evening, they can occur at any time, often with little or no warning.**

    The key to survival is advanced planning. All members of a household should know where the safest areas of the home are. Identify interior bathrooms, closets, halls, and any basement shelter areas. Be sure that every family member knows to move to such areas at the first sign of danger—there may be only seconds to act. Also have a tornado emergency plan at work, and encourage area schools to form a tornado plan and conduct drills.

    Tornadoes occur in many parts of the world and in each of the fifty states. No area is more favorable to their formation, however, than the continental plains and the Gulf Coast of the United States during April, May, and June. Tornadoes are least frequent in the United States during the winter months, although damaging tornadoes can develop at any time of year.

Tornadoes are perhaps nature's most spectacular events. Thousands of people across the United States spend their vacations in Tornado Alley, bouncing down dusty back roads and cruising interstates in hopes of seeing and photographing a tornado. Few experience frequent success, because it takes a great deal of weather knowledge to discern when and where conditions will be right for nature to spawn one of these awesome phenomena.

Not all violent storms, however, require so much skill to forecast and find. There are some that no one chases—they come to you and stay for days. We're talking about winter's answer to the summertime tornado show: the blizzard. If one heads your way, you won't need special chasing equipment. You'll have plenty of time to see it up close and personal.

The Aftermath of a Tornado in Salina, Kansas

Blizzard!!

January 11, 1975, brought one of the worst blizzards in history to Minnesota. It was locally dubbed the "Storm of the Century." In Duluth, two feet of snow fell, the pressure dropped to 28.55 inches. As the storm passed, winds gusted to 80 mph across the state, whipping snow into 20 foot drifts and causing wind chills to bottom out at –80F. When all was said and done, 60 people lost their lives, hundreds were injured and damage estimates were placed upward of $15 million.

How It Happens

Such damaging weather gets its start when a deep low pressure system moves across a region and dumps considerable amounts of snow. In some cities, the snowfall can be enough to close businesses, snarl traffic, halt airport operations and interfere with railroads. But the blizzard itself hasn't yet begun.

The storm passes and the true snowfall usually draws to a close. Meanwhile, cold air and strong pressure gradients tailing the storm invade the area causing temperatures to plummet and

Worst Winter Storms in the United States

The Blizzard of 1888 was the first major blizzard that was accurately documented for history. Since then, there have been other serious blizzards but none more deadly. Notice that 1888 was the first recorded blizzard before the widespread use of electronic communication tools. Certainly, today, we benefit greatly from the advance warning that would have saved many lives in 1888.

March 11–14, 1888
East Coast: 5 feet of snow, 400 deaths.

January 27–29, 1922
Washington, D.C., area: 100 deaths.

November 11–12, 1940
Northeast and Midwest: 144 deaths.

December 27, 1947
New York City area: 80 deaths.

November 26, 1950
East Coast: 250 deaths.

February 15–16, 1958
Northeast: 171 deaths.

January 29–31, 1966
East Coast: 165 deaths.

December 12–20, 1967
Southwest: 51 deaths.

January 28–31, 1977
Buffalo, New York area: 25-foot drifts, 29 deaths.

February 6–7, 1978
New England: 339 homes destroyed, 29 deaths.

November 28–December 1, 1985
Midwest: 26 deaths.

January 22, 1987
East Coast: 37 deaths.

March 13–14, 1993
Eastern United States: Hurricane-force winds from Cuba to Canada, record low pressure, and 15-foot snow drifts; 184 deaths.

January 6–10, 1996
Eastern United States: Followed by severe flooding from rain and snowmelt, $3 billion in damage, 187 deaths.

---

winds to pick up. Over time, the winds get worse, and within hours, vast amounts of snow are whipped up hundreds of feet into the air, turning an otherwise clear day into a vicious storm of blowing snow. It often is impossible to tell whether the snowstorm itself has stopped or not.

If enough free, powdery snow is carried into the air, a condition known as a "whiteout" may occur, where visibility is restricted to 10 feet or less. This completely encloses the observer in a disorientating whiteness, which can be so intense that sensations of gravity seem to disappear. The worst of whiteouts are encountered in polar regions and often are caused by ice crystals as well as blowing snow. Although they can last for days, the storm is sometimes only about 10 feet deep.

A Look at a Blizzard

This chart shows the weather situation on January 10, 1975, the morning before the Great Blizzard. A polar front extended across the Mississippi Valley, with a low and occluded front over Kansas. Deep moisture and warm air was moving across the southeast United States, fueling the surface low with water vapor and instability.

THE WEATHER SOURCEBOOK

Snow piles in Sandy Creek, New York, January 1977

Looking aloft, a deep upper-level trough was positioned over the Great Plains, routing a strong jet stream from Texas up to Illinois. Strong dynamics associated with the jet stream resulted in a large region of horizontal upper-level divergence over Iowa and Minnesota.

When the weather observer at Kirksville, Missouri, measured the atmospheric pressure that morning, it was about 29.05 inches. But the divergent motions occurring aloft were expanding the air horizontally over the region, spreading the air outward and away in all directions and causing the total mass of air at any given point to decrease. At weather stations across the region, mass was decreasing in the vertical columns above the stations. Therefore, the pressures were decreasing. In only 3 hours, the pressure at Kirksville had dropped by a fifth of an inch!

The falling pressures deepened the surface low rapidly as it moved northward with the jet stream into Minnesota. Surface winds intensified and converged into the low, trying to fill it up. But a circulation had developed—air was converging at the surface and had been for some time diverging aloft. This meant that surface air had to rise. When combined with generous amounts of moisture spiralling into the circulation from the southern United States, precipitation, clouds

CHAPTER 3: VIOLENT WEATHER

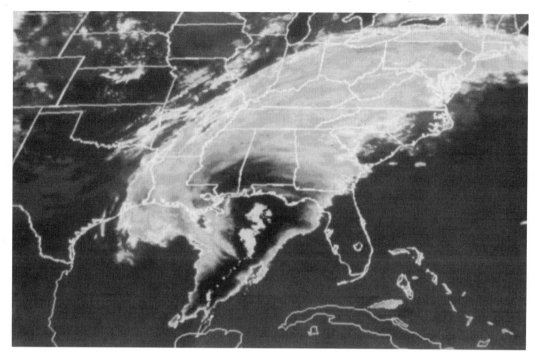

Satellite Imagery of The Blizzard of '93

and bad weather resulted. Up to two feet of snow covered the ground in Minnesota and surrounding regions.

The mere intensity of the surface low caused a strong pressure gradient in its vicinity, causing 30 and 40 knot winds to blow across the northern United States. And where these winds gusted across the snow cover, the stage was set.

The blizzard howled across Minnesota.

—By Tim Vasquez of the AAWO. Reprinted with permission from the *American Weather Observer*.

Facts Behind the Fiction

A lot of the classic weather lore is based solidly on verifiable weather facts. Obviously people long ago adopted some fairly scientific folklore that could reliably predict the weather. Here are a couple of examples:

When the cow scratches her ear,
It means a shower is near.
But when she thumps her ribs with her tail,
Expect thunder and lightning and hail.

The hairs inside a cow's ear respond to the changes that come before rain (low atmospheric pressure and increased humidity) and may cause her to scratch them. Before a violent thunderstorm, static charges of electricity can cause a cow's hair to stand out. To relieve this discomfort, a cow may continuously brush herself with her tail. Found generally in the South, Scotland, and England, the following rhyme actually is a fair predictor of weather:

If February brings drifts of snow,
There will be good summer crops to hoe.

During a very cold, snowy winter, some weed seeds are damaged, many insect larvae killed, and an abundance of water is stored in the ground. Later, the scarcity of weeds and insects that might otherwise damage crops, in addition to the abundance of water available to young plants, provides good conditions for crop growth and increases the chances for a rich summer harvest. Found in the mountainous regions of the South, the following poem is a reliable predictor of crop yields:

When chickens scratch together,
There's sure to be foul weather.

Since chickens' feathers trap air (as insulation), they quickly feel the changes in air pressure, moisture, and temperature that come before a storm. These changes may be sufficient to make them restless, move about more, or scratch together to keep warm. Found in the South and in New England, this poem is a good weather indicator.

CHAPTER 3: VIOLENT WEATHER

Products Related to Violent Storms

Since violent storms affect so many people in such a large magnitude, it's only natural to find that equipment has been designed to help warn of impending destructive storms. Here we've listed products that can help warn you of tornadoes and track hurricanes so you can prepare properly for their arrival.

AllSky Lightning Storm Detector P-10
Airborne Technology Corporation
240 Bear Hill Road
Waltham, MA 02154
(617) 890–8381
fax: (617) 890–7411

The P-10 is a storm-warning system that uses electro-optical detection to alert you of approaching dangerous storms. Its adjustable range settings allow you to choose how early a warning you want. When an approaching storm reaches your preset range, an alarm sounds that can be heard up to 500 feet away; and there is a provision for a second remote alarm that can provide an even wider alert area. The P-10 is a component system composed of a base unit and a removable, handheld unit called the M-10. The M-10 operates on batteries that are recharged by the base unit and that can provide remote power for up to two days.

The P-10 can sense storms that are hundreds of miles away. It's so sensitive, it can detect intracloud lightning that occurs ten to thirty minutes before hazardous cloud-to-ground lightning. The P-10 costs $1,200, and the M-10 is available separately for $600.

Hurricane Tracking Chart
Blue Hill Meteorological Observatory
P.O. Box 101
East Milton, MA 02186
(617) 698–5397

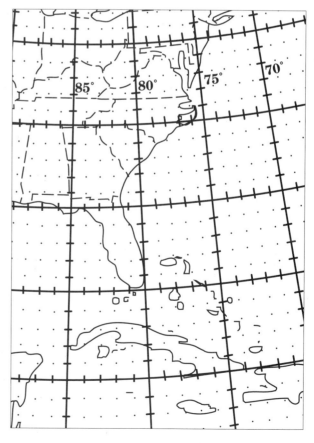

Blue Hill stocks a hurricane-tracking chart that includes a 16- by 20-inch map of North America along with instructions on how to plot a hurricane's course using longitude and latitude. There is also a hurricane-survival checklist and a list of current hurricane names. The 1988 version of the chart includes information about the 1938 New England hurricane. This nonprofit observatory offers weather-related publications, including a quarterly bulletin for members. Membership is $10 a year. Write for a free brochure.

Stormfax

Stormfax Weather Services
P.O. Box 684
Bryn Mawr, PA 19010
(800) 88–STORM
fax: (215) 896–1668

Stormfax serves business, industry, and private individuals from coast to coast with a facsimile weather service. For example, in August 1993, when Hurricane Emily was threatening the vacation beaches along the Atlantic coast, Stormfax transmitted regular, automatic updates to customers that included a table listing the storm's position, movement, wind direction, wind speed, and barometric pressure. A brief written forecast was included with each table. Even if you are completely out of touch with local news sources, Stormfax will alert you by fax, ensuring that an approaching hurricane won't catch you off guard.

Weather Warnings by Fax

Accu-Weather, Inc.
619 West College Avenue
State College, PA 16801
(814) 234–9601, extension 400
fax: (814) 238–1339

For a $49 annual subscription fee and $1.99 per notice, you can automatically have official severe-weather notifications faxed to you immediately as they are issued by government agencies. You can select a choice of watches, warnings, advisories, or special statements; and you can specify the events for which you want to receive notification: tornadoes, severe thunderstorms, hurricanes, tropical storms, flooding, high winds, heavy rain, snow/ice, and dense fog. The service delivers to you twenty-four hours a day, 365 days a year.

National Severe Storms Laboratory

1313 Halley Circle
Norman, OK 73069
(405) 360–3620

This extension of the National Weather Service is the ultimate source for information and historical data on violent storms. It's our leading research center on thunderstorms, tornadoes, strong winds, and lightning. The research is aimed at increasing flight safety and learning how to predict storms more accurately so that severe-storm warnings will be more timely.

Pet Tornado

TAZCO
11675 Valewood Drive
Victorville, CA 92392

You don't have to spend a fortune and your entire vacation driving dusty back roads in Oklahoma or Kansas to see a tornado. Here's a toy that really makes pet rocks look boring, except you can't actually pet this "pet." The Pet Tornado is a small jar that contains a secret formula that generates a tiny tornado when you shake the jar. When shaken properly, a tornado forms, swirls harmlessly for a while, then—as all good little tornadoes should—disappears into the "clouds." The hand-powered version, priced at $3.95 (plus $1.00 shipping and handling), was the only model available at this writing, but by now they also will have a three-speed, battery-powered version.

Chapter 4
Atmospheric Pressure

If on some days you feel that the weight of the world rests on your shoulders, consider these facts: When you walk, you are carrying a burden of a couple of tons, and when you lie down, you could be supporting 16,000 pounds! Why? Because the air in the atmosphere is pulled down to Earth by gravity, which means it has weight. A column of air 1 square foot weighs about one ton, and you've got a 2-square-foot column of air above you when you walk. You're not crushed under this burden, though, because air follows the physical principles of fluids and presses equally in all directions, so you've got a balancing pressure that keeps you from being flattened.

We can play tricks with air pressure and create *barometers*, devices that respond to the weight of the air above them. Barometers help us measure the air's weight, which we call air pressure. Air pressure varies according to daily conditions and has a large impact on the day's weather. First let's see why the weight of the air varies, then we'll find what we can gain by learning the significance of the variances.

Pressure Changes: Causes and Effects

The temperature of air affects its density. When air is cold, its slower molecular movement keeps it more tightly packed so more air molecules fit into a given space. As air is heated, the increased molecular movement causes the molecules to fly apart, so fewer fit into a given space. Temperature, therefore, alters the total weight of the air column being measured. A column of air weighs more on a cold winter day than it does on a hot summer day. Thus, variations in air-pressure readings would be observed even if the only change were in its temperature.

An air column's moisture content also will affect its total weight. Air with high humidity is less dense than dry air; therefore, a correction would have to be made for a pressure reading to be perfectly accurate. Humidity affects air pressure to a lesser extent than does the temperature factor, however, and often is ignored by home weather-station users.

When local conditions cause the pressure in an air mass to drop, it creates a low-pressure area that meteorologists call a *low,* or *cyclone*. High-pressure areas are called *highs,* or *anticyclones*. Nature does not tolerate out-of-balance conditions, so the physical laws of equilibrium cause air from high-pressure areas to flow toward areas of lower pressure. But equilibrium cannot be established instantly, and while there is an imbalance, we have plenty of time to see the effects that a low or a high can have on local weather.

We have learned that high-pressure areas exist because they have a greater amount of air stacked on top of them than other areas. The increased weight compresses the air at the bottom. Compression causes heating, so the temperature of the air rises above the saturation point, causing clouds to evaporate.

In low-pressure areas, air at the bottom is not compressed as much. Air currents tend to flow upward. As air rises, it cools and may reach its saturation point, forming clouds.

Nature is constantly trying to establish air-pressure equilibrium, but other factors keep that from occurring. Instead, our planet is covered with a continually shifting pattern of lows and highs. These ever-changing pressure systems determine our weather and can have a dramatic impact on it. The larger the pressure change, the more dramatic its effect. The next section provides a close-up look at what kind of weather a record-breaking pressure system can generate, followed by an explanation of how such a deep, low-pressure system is able to counteract the natural laws of equilibrium.

Pressure Gone Amok: "The Storm of the Century"

On March 12–15, 1993, a storm now called "The Storm of the Century" struck the eastern seaboard. Comparisons to the Blizzard of 1888 are now being made. To set the context for understanding the record-setting magnitude of the 1993 storm, here are a few facts from the '88 storm:

- 400 people killed
- 50 inches of snow in Saratoga Springs, New York
- 48 inches of snow in Albany, New York
- 22 inches of snow in New York City
- Snowdrifts over the tops of houses from New York to New England
- 80 mph wind gusts common

Although the 1888 storm probably was more severe in the Northeast and New England, it did not affect the entire eastern seaboard to the extent that the 1993 storm did. Following are some highlights of the information compiled about the 1993 storm:

1. A spokesman for the National Weather Service's (NWS) special studies branch said that the volume of water that fell as snow may be unprecedented. The NWS office at Asheville, North Carolina, reported a snow/water ratio of 4.2 to 1 from core samples of new snow. This equated to more than 5 inches of liquid equivalent precipitation in some areas. Areas north of Asheville that reported up to 4 feet of snow probably received drier snow with similar liquid equivalent amounts. Due to the weight of the heavy snow, damage to trees and some buildings was extensive. Polk County, North Carolina, reported 99 percent of its electrical customers without power at one point during the storm.
2. At least 243 deaths were attributed to the storm, plus 48 missing at sea. This is more than three times the combined death toll of 79 attributed to hurricanes Hugo and Andrew.
3. Thousands of people were isolated by record snowfalls, especially in the Georgia, North Carolina and Virginia mountains. More than 100 hikers were rescued from the North Carolina and Tennessee mountains. Curfews were enforced in many counties and cities as states-of-emergency were declared. The National Guard was employed in many areas, especially in the North Carolina mountains.
4. For the first time, every major airport on the East Coast was closed at one time or another by a single storm.
5. Thousands of roof collapses were reported due to the weight of the heavy wet snow. More than 3 million customers were without electrical power at one time due to fallen trees and high winds.

CHAPTER 4: ATMOSPHERIC PRESSURE

6. At least 18 homes fell into the sea on Long Island due to the pounding surf. About 200 homes along North Carolina's Outer Banks were damaged and may be uninhabitable.
7. Florida was struck by at least 27 tornadoes and at least 15 counties were struck. Twenty-six deaths in Florida were attributed either to the tornadoes or other severe weather. A 9-foot storm surge was reported in the Apalachicola area. Preliminary damage estimates ran as high as $1 billion. Up to 6 inches of snow fell in the Florida panhandle.
8. Three storm-related deaths were reported in Quebec and 1 in Ontario. Three deaths occurred in Cuba (Havana was blacked out), and a tornado left 5,000 people homeless in Reynosa, Mexico, near the Texas border.

Observational Data from March 12–15, 1993

Record low sea-level pressures:

28.38 inches in White Plains, New York
28.43 inches in Philadelphia, Pennsylvania
28.43 inches at JFK Airport, New York
28.45 inches in Dover, Delaware
28.51 inches in Boston, Massachusetts
28.53 inches in Augusta, Maine
28.54 inches in Norfolk, Virginia
28.54 inches in Washington, D.C.
28.61 inches in Raleigh-Durham, North Carolina
28.64 inches in Columbia, South Carolina
28.73 inches in Augusta, Georgia
28.74 inches in Greenville-Spartanburg, South Carolina
28.89 inches in Asheville, North Carolina

Highest recorded wind gusts:

110 mph in Franklin County, Florida
110 mph on Mount Washington, New Hampshire
101 mph on Flattop Mountain, North Carolina
98 mph in South Timbalier, Louisiana
92 mph on South Marsh Island, Louisiana
90 mph in Myrtle Beach, South Carolina
89 mph in Fire Island, New York
83 mph in Vero Beach, Florida
81 mph in Boston, Massachusetts
71 mph at La Guardia Airport, New York

Snowfall totals:

50 inches on Mount Mitchell, North Carolina
(14-foot drifts)
44 inches in Snowshoe, West Virginia
43 inches in Syracuse, New York
35 inches in Portland, Maine
35 inches in Lincoln, New Hampshire
30 inches in Beckley, West Virginia
29 inches in Page County, Virginia
27 inches in Albany, New York
25 inches in Pittsburgh, Pennsylvania
24 inches in Mountain City, Georgia
20 inches in Chattanooga, Tennessee
19 inches in Asheville, North Carolina
17 inches near Birmingham, Alabama
(6-foot drifts)
16 inches in Roanoke, Virginia
13 inches in Washington, D.C.
9 inches in Boston, Massachusetts
4 inches in Atlanta, Georgia

Record low temperatures (some records for March):

−5° F in Elkins, West Virginia
−4° F in Waynesville, North Carolina
1° F in Pittsburgh, Pennsylvania
2° F in Asheville, North Carolina, and Birmingham, Alabama
6° F in Knoxville, Tennessee
8° F in Greensboro, North Carolina
9° F in Beckley, West Virginia
11° F in Chattanooga, Tennessee, and Philadelphia, Pennsylvania

15° F in New York–JFK and Washington, D.C.
17° F in Montgomery, Alabama
18° F in Columbia, South Carolina, and Atlanta, Georgia
19° F in Augusta, Georgia
21° F in Mobile, Alabama
25° F in Savannah, Georgia, and Pensacola, Florida
31° F in Daytona Beach, Florida

The data lists above were taken from decoded surface observations for the period March 12–15, 1993. The two most damaging characteristics of the storm were the heavy, wet snow and the high winds, including tornadoes. A complete file of all observations (56,090 observations—4.4 megabytes) during this period from latitude 20°N to 50°N and longitude 65°W to 95°W is available as shown below. Additional data, including elements such as wave heights, from these observations are available from the National Climatic Data Center's Research Customer Service Group upon request in these formats:

- 4 diskettes (3.5-inch, high density)
- Magnetic or cartridge tape
- INTERNET via FTP:
  - open 192.67.134.72 or open hurricane.ncdc.noaa.gov
- Login is: anonymous
- Password is: guest
- You are now logged onto a UNIX workstation. Enter "help" if you'd like a list of available commands.
- To move to the correct subdirectory, enter: cd /pub/upload/blizzard
- To get a copy of the data, enter: get storm.txt destination
  (Destination is your output location and name. For example, "get storm.txt c:storm.txt" copies to hard drive c.)

—Compiled from information downloaded from the WX-TALK forum on the University of Illinois Weather Machine.

We know that storms are born out of low-pressure systems. Yet when you consider the laws of equilibrium, it may not seem possible for a pressure system as deep as this one to develop. You might expect that as pressure dropped in a region, the natural course of physics would cause higher pressure to flow in and establish a balance. Sometimes, though, as pressure begins to drop, other factors contribute to a deepening of the already low-pressure area. If those contributing factors are stronger than the balancing forces of nature, the incoming high pressure cannot feed the low as fast as it deepens.

It takes an unusual combination of weather events to create the conditions that produce a low deep enough to spawn a storm. It takes a truly rare combination of weather events to spawn a record breaker. What rare combination converged to create such a massive storm as the March 1993 "Storm of the Century"? Here's a simple explanation from weather expert Phil Leith, describing how an already low-pressure system can deepen instead of being restored to equilibrium.

How "The Storm" Was Born

It seems that a 190-knot jet stream came roaring through at the 200mb level directly over the storm center. This caused enhanced 'mass removal' at the surface and deepened an already strong storm to immense proportions. The occurrence of one of the two systems is seldom. Getting the two to coincide is rare. Hence, "The Storm of the Century."

CHAPTER 4: ATMOSPHERIC PRESSURE

Here's a diagram that illustrates the conditions:

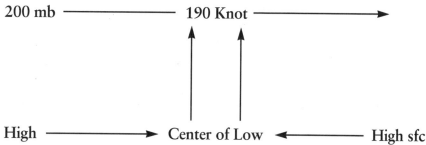

It's all a question of conservation of mass. Suppose you've already got a weak low-pressure system, an area with pressure relatively low compared to its surroundings. Air tries to flow from high pressure to low pressure to equalize the pressure.

We have low pressure, air flows into the area from the surrounding high pressure areas and it fills up and everything is equal, but nothing lasts forever. When the air flows into the area, it comes up against air coming from the opposite direction. That gets the ball rolling.

Then, Crash! It can't go down, there's ground down there. The only way to go is up. And it does until the mass is equalized and then it stops because there's no high or low pressure—it's all the same now.

Enter a strong jet stream aloft. Now you have air flowing into the low, rising and being carried away by the strong upper level winds. So it never fills up and air keeps flowing in.

Now suppose that the jetstream is carrying off more air than is flowing into the low. You get less mass in the column of air where the low is—causing lower pressure—which means more air has to flow in to replace it so it comes in faster, which means when it gets there it rises faster and the air cools and the water condenses as it rises and precipitation falls. The stronger this process is, the more cloud and precipitation potential increases.

The models all take things like kinematics, thermodynamics, the rotation of the earth—everything we've got a theory about—into account. Then we simply ignore things that shouldn't matter much and guess at a few things we don't know, tweak the model where it shows flaws and after all that it's pretty amazing that they work as well as they do! Think—we have about 100 upper air stations across the United States. That's some pretty sparse data to come up with numbers as close as we do.

So, we fed the model some initial data and it showed a low developing in an extremely favorable area, and the model extrapolated out what should happen according to the "laws" of physics that we felt applied enough to include in the model. That showed a very intense storm developing because of all the support it had. Lots of moisture, good upper air support and baroclinic instability running amok (high temperature contrast plays a big part here). And it turned out the model was basically right, as it often is—especially in the 48-hour arena. The forecasters believed it, sent watches and warnings out, beefed up staff—the rest is history.

—By Phil Leith, from WX-TALK, March 17, 1993

Pressure-Measuring Instruments

In 350 years we have not found a better way to measure atmospheric pressure than the method accidentally discovered by Evangelista Torricelli in 1644. He no doubt never would have guessed that his invention would serve as the foundation of a vast high-tech weather network that employs man-made satellites and high-speed supercomputers.

Torricelli created the barometer merely as part of an experiment to prove his theory that air had weight. Afterward he noticed that the height of the mercury remaining in the barometer varied daily. Before long, Torricelli observed that the mercury level seemed to correspond to local weather conditions: As the mercury level dropped, the weather deteriorated; as it rose, the weather improved. Thus was born the barometer—the "weather glass" that remains the most reliable of many atmospheric-pressure measuring devices.

Today, the most widely used, if not the standard, barometric instrument is the aneroid barometer. *Aneroid* means "without liquid," which is a terrific quality when the liquid that it is without is the highly toxic element mercury.

Still, the mercury barometer is the most stable and accurate type, so we always will need it, if only to calibrate the more convenient, alternative models. Using these other models to obtain reliable, accurate pressure readings requires a surprising amount of care. Let's now learn how to properly use, care for, and calibrate the popular barometer models available today.

Barometer Care

Temperature and sunlight can affect a barometer's readings. For greatest accuracy, the barometer should be located outdoors—but not exposed to the elements or to direct sunlight.

The process of measuring air pressure is different from gathering any other weather data: To be significant, it must be considered in a relative sense. Temperature isn't like that; 100° F is hot both in Denver and in New York. However, since atmospheric pressure is the measure of the weight of the air above a given location, readings vary with altitude.

A pressure reading that might indicate fair weather in mile-high Denver would indicate severe weather conditions in near-sea-level New York. Why? The column of air above a mountain location is much shorter than the one over a sea-level location. Since the shorter column weighs less, it produces overall lower pressure readings than the one at the sea-level location, even in fair weather conditions. Thus, for barometric readings to be meaningful for predictions, we first must establish a baseline for each location being measured.

Barometric Elevation Correction

Correct adjustment for a barometer's elevation above sea level is absolutely crucial to meaningful pressure readings. To adjust for the variation in the height of the air column being measured at different elevations, barometric-pressure readings customarily are corrected to their equivalent reading at mean sea level. A rule of thumb is to adjust barometer readings by 1 inch of mercury for each 900 feet above mean sea level.

First you must determine the height above sea level at the barometer's location. This value can be obtained by using a topographical map of your area, available from the U.S. Geological Survey in Washington, D.C., as well as through local distribution channels throughout the United States. Use the map carefully, because for accurate readings you need to know your elevation above mean sea level within 5 feet.

To achieve greatest accuracy, you can use one of the thousands of benchmarks that were erected by the U.S. Geological Survey early in this century. These benchmarks are small stone markers that state the marker's exact location and elevation. If you cannot find one in your area, ask your state, county, or municipal engineer to help you establish your elevation using the "1929 Datum Plane Adjustment for Mean Sea Level."

If you can determine a precise elevation, you may want to use a precise correction factor. You can achieve professional-quality accuracy by constructing a table for your location, in inches of mercury, using formulae or master tables such as the ones found in *The Manual of Barometry*, Volume 1 (1963), published by the U.S. Government Printing Office in Washington, D.C. This book will help you follow the same guidelines used by the National Weather Service. To construct a correction table using metric readings (millibars), obtain the *Observer's Handbook*, Second Edition (1956), published by the British Meteorological Office. To order a copy, write to Her Majesty's Stationery Office, care of Unipub, 4611 South Assembly Drive, Lanham, MD 20706; or call (800) 274-4888.

If your elevation is less than 500 feet above mean sea level, you will get accurate barometer measurements by correcting your raw readings according to the "Standard Atmosphere" model. This model is used in aviation to correct aircraft altimeters. It is based on a theoretical atmosphere with a mean sea level reading of 29.921 inches of mercury, with a standard lapse rate applied for elevations above sea level.

Barometric Temperature and Moisture Correction

While elevation adjustments are crucial to meaningful barometric readings, temperature correction is not nearly as important. For elevations of 50 feet or less above mean sea level, the correction is nil. The National Weather Service standard practice ignores temperature variations on barometric readings up to 500 feet above mean sea level, since they will be small. Above 500 feet, however, temperature variations can significantly alter raw pressure readings, so you need to apply a correction factor for meaningful readings.

Barometric readings are not customarily corrected for variations in humidity.

Barometer Maintenance and Testing

Any delicate measuring device should be checked regularly to ensure that its readings are accurate. The type of barometer used will determine the frequency of checks required. Mercury barometers require the least maintenance and calibration and are highly reliable. The moving parts in aneroid barometers require greater vigilance to ensure accuracy. They can be subject to changing errors caused by mechanical wear of the internal components. The evacuated capsule—a sealed chamber with a partial vacuum that is the heart of an aneroid barometer—needs particular care. Even the latest, high-tech electronic barometers with no moving parts need to be checked for instrumentation drift that can creep in over time and be subject to temperature variations.

Aneroid and electronic barometers usually are checked against calibrated and certified mercury barometers on a regular basis. Observers at National Weather Service or Federal Aviation Administration stations check their instruments every six hours as part of their normal observation routine. With some of the highly accurate, precision aneroid barometers, observers need to check them only once each week.

Since it's unlikely that you will have ready access to a certified mercury barometer, follow these steps as an alternate means of getting an occasional barometer check:

1. Use the closest possible National Weather Service or Federal Aviation Administration observation station.
2. Make sure that the pressure system on the day of your check is slow-moving and stable.
3. Make sure that the temperatures between your station and the comparison station are close.
4. If there is a significant temperature variation, use correction tables to account for the difference.
5. Record any difference in the two readings and apply it to readings from your barometer until your next check.

If you use aircraft altimeter settings as your comparison guide, you'll be benefiting from the temperature corrections that are built into the Standard Atmosphere. Altimeter settings will smooth out the pressure-reading variations you might get from temperature extremes. Since these will require less work than applying temperature corrections yourself to standard pressure readings, you might be more likely to make biweekly or monthly barometer checks. More frequent checking will reduce the effects of varying temperatures on your sea-level pressure readings and minimize their variations from official readings.

Atmospheric-Pressure—Related Products

Now that you know how to use one properly, it's time to go shopping for your barometer. In this section we present the collection of good barometers promised earlier. Try out whichever one seems to suit you best, using what you've just learned about placement, reading, and periodic comparison testing.

DIGITAL BAROMETERS

Nimbus Barometer
Sensor Instruments Company, Inc.
41 Terrill Park Drive
Concord, NH 03301
(800) 633–1033
in New Hampshire: (603) 224–0167

Want to know the pressure four hours ago? How about thirty-five days ago? The Nimbus, with its long-term memory, will give you the answer for any hour over the last thirty-five days. You can read its answers on the large LCD display in inches of mercury, millibars, or kilopascals. It also has a high/low memory so you can track weather systems and storms as they pass. You can have your choice of a cherry or oak case. The unit is powered by four "C" cell batteries and includes a three-year warranty. The list price is $350. An optional RS-232 computer interface is list-priced at $100.

Automatic Weather Forecasting Barometer
The Sharper Image Catalog
(800) 344–4444

This unique instrument is from Oregon Scientific. Taking advantage of digital technology, it goes beyond being merely a digital readout. Its large LCD readout displays indoor relative humidity (between 10 and 98 percent); temperature (between 23° F and 131° F); time; date; and barometric pressure in a six-line bar graph that shows the current reading plus the reading for each of the last six hours. Cursor keys let you cycle the display through the readings for the last twenty-four hours.

What is unusual about this automatic barometer is its forecasting computer, which analyzes your barometric history and displays icons—a sun, a sun with clouds, clouds only, and clouds with rain—that tell you the coming weather. The setup mode is easy and lets you select temperature readings in degrees Fahrenheit or Centigrade, pressure in inches of mercury or millibars, and time in a twelve-hour or twenty-four-hour display; and lets you enter your altitude above sea level. It is priced at $129.95 plus shipping and handling.

CHAPTER 4: ATMOSPHERIC PRESSURE

MERCURY BAROMETERS

Wall-Mounted Mercury Barometers
Robert E. White Instruments, Inc.
34 Commercial Wharf
Boston, MA 02110
(800) 992–3045

For the highest standards of accuracy in air-pressure measurements, many meteorologists choose a mercury instrument. White offers two models. The Nova is a school-grade model that has an aluminum cylinder, friction vernier, and thermometer; is calibrated in inches and millibars; and can be used at elevations as high as 10,000 feet above sea level.

A higher-grade National Weather Service model is more rugged and accurate with brass cylinder, rack-and-pinion vernier, and thermometer. This instrument is also calibrated in inches and millibars but is usable only to 3,000 feet above sea level. Accuracy on this model is ±0.3 millibar. Both models are 42 inches tall and come with instructions and correction tables.

Classic Admiral Fitzroy Barometer
Robert E. White Instruments, Inc.
34 Commercial Wharf
Boston, MA 02110
(800) 992–3045

The father of the U.S. Coast Guard designed this complete weather-forecasting system, and you can have your own with this classic reproduction. Admiral Fitzroy used this instrument on his five-year voyage in command of the HMS *Beagle* with Charles Darwin, which resulted in the publication of Darwin's *Origin of Species*. This historic instrument includes a mercury barometer, a thermometer, a storm bottle, and a clock. Elegant front panels aid in interpreting the comprehensive information it provides. It's built in a polished mahogany case nearly 4 feet tall and includes illustrated documentation explaining its use and history. This is a great gift idea, priced at $595.

ANEROID BAROMETERS

For the Perfectionist
Robert E. White Instruments, Inc.
34 Commercial Wharf
Boston, MA 02110
(800) 992–3045

The Terra barometer may be the most accurate you can buy. It has a huge range (25.80–31.00 inches of mercury), graduated in 0.01-inch increments (and in millibars) displayed on a white 5-inch face in a very rugged, chrome-plated brass case. Its "zero-gauging" mechanism—two opposing aneroid sensor capsules that eliminate friction and lag—gives it extraordinary accuracy. This superaccurate model should satisfy the most demanding perfectionist. It's not cheap; call White for pricing.

For the Budget-Minded
American Weather Enterprises
P.O. Box 1383
Media, PA 19063
(215) 565–1232

This is a highly accurate, low-cost aneroid barometer from the German company Lufft. This model, the 2179, is a good trade-off between price and accuracy. Its 6-inch brass case makes it rugged and reliable; its 5-inch white face (marked in both inches and millibars) makes it sensitive and accurate; and its wide range (from 27.5 to 31.5 inches) makes it versatile. It is designed for wall mounting and is priced at $135.

RECORDING BAROMETERS

Barographs
Robert E. White Instruments, Inc.
34 Commercial Wharf
Boston, MA 02110
(800) 992–3045

White offers a wide variety of barographs, but two stand out at excellent bargains. One is a low-cost model from Taylor that records for seven-day periods and, list-priced at $384, is a good choice for

amateurs or those on a budget. It has a sturdy, black plastic base with a clear plastic cover and is powered by 110 volts AC.

They also have a Maxant mahogany model that records for either seven-day or thirty-one-day periods and lists for $600. It's an attractive desk piece that has durable brass hardware and a quartz movement that operates on an "AA" battery (included). Both units include charts, pens, and instructions and are covered by warranty.

BAROMETERS ON-THE-GO

Here are some good travel barometers. Of course you won't get professional-quality precision unless you come across a surveying benchmark while you're out. Their purpose, though, is to help you get early warning that weather conditions are about to change. So, while not useful for establishing a database line back home, they can be valuable aids during extended outdoor activities where your main interest lies in tracking atmospheric-pressure changes.

A Leather-Cased Brass Travel Barometer
The Nature Company
Catalog Division
P.O. Box 188
Florence, KY 41022
(800) 227–1114

This sensitive travel barometer measures air-pressure changes and warns of impending shifts in the weather. It adjusts to altitudes up to 4,500 feet. The Nature Company also sells a matching compass, always handy for travelers who would use the barometer and an essential safety tool for hikers. Both include lined, leather cases with handy straps. Together these pocket instruments create a wonderful gift set that would be indispensable to wilderness trekkers and fun at any outdoor occasion. The barometer costs $75, the compass, $69.

Digital Altitude and Temperature
American Weather Enterprises
P.O. Box 1383
Media, PA 19063
(215) 565–1232

The Ultimeter Model 3, made by Peet Incorpo-

CHAPTER 4: ATMOSPHERIC PRESSURE

rated, displays barometric pressure; altitude above sea level (up to 16,400 feet in 3-foot increments); and temperature, either in degrees Fahrenheit (–20° F to 120° F) or Centigrade. Pressure measurements are in 0.01-inch increments and are temperature compensated to –27 feet. It operates on three "AAA" batteries and has an optional NiCad battery pack with AC/DC charger. The list price is $209.95.

Casio Barometer Watch

Wernikoff's Jewelers
2731 N. Milwaukee Avenue
Chicago, IL 60647–1388
(800) 932–8463

This electronic, digital watch/barometer from Casio, Model #WW2, seems to be the ultimate in compactness and portability in barometers—you literally can take it everywhere you go. In the barometer mode, this watch reads from 610 to 1050 millibars in 1-millibar increments. It has a computer memory that stores pressure data for the past eighteen hours. As an altimeter, the WW2 can be used from sea level to 4,000 meters, reading in 5-meter increments, and it includes an altitude alarm. It even works below sea level, permitting you to measure water depths up to 30 meters, and includes a depth alarm. It comes with a one-year warranty, a two-year battery, and is priced at $89.98 plus $5.00 shipping and handling.

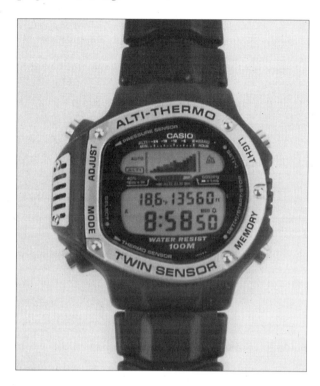

Chapter 5
Wind

An imaginary wind can carry a little girl in Kansas to a fantastic dream world. A real-world wind can bring in a deadly fog or transport clouds with rains that devastate entire communities. Wind makes phenomena like acid rain possible when it carries the pollution from factories into pristine wilderness areas many miles away. The same wind also can bear the seeds of the trees in a poisoned forest and ensure that a new generation of trees will grow someplace else, perhaps far enough away to survive and be free of the ravages of humankind that may have killed their forebears.

Wind is simply air in motion. You can see it sway trees, move leaves, and ruffle your hair. You can feel it on your face. Small wind patterns can circulate in the corners of buildings, picking up leaves and trash in a swirl. Huge wind patterns can span vast areas, picking up houses and cars.

Winds are named for the direction of their origin. Thus, a "north wind" blows *from* the north. When wind direction changes, you usually can count on changing weather. For example, in the Northern Hemisphere, a south or west wind often brings warm or mild, wet weather; and a north or east wind often brings colder, drier weather, especially in the winter.

Wind generates and controls our global system of weather, which sweeps the planet in well-established patterns. The patterns are set in motion by the differing strengths of the sun's rays that penetrate different areas. Air expands as it's heated, becoming lighter. The air above hot areas expands, rises, and spreads out. The air above cold areas condenses, sinks, and becomes heavier and more tightly packed. These differences in weight and action set the wind blowing as nature constantly attempts to restore equilibrium.

In very hot tropical regions, the air expands and rises to such a great extent that little wind blows on the surface. Sailing ships have been becalmed for weeks at a time in the equatorial tropical regions known as the *doldrums*. This rising air flows toward the colder polar regions. At about 20° to 30° of latitude, both north and south, this air becomes cool enough to begin sinking into the regions we call the *subtropics*, where the pressure generally is high, the winds calm, and the weather clear. Over land, these regions contain the world's deserts; over the oceans, they are called the *horse latitudes* (which will be explained later in this chapter).

The laws of equilibrium rule the weather in this region. The air near the surface in the subtropics splits and moves in two directions, flowing back toward the equator to fill the void left by the hot, rising air in the doldrums. This steady flow is affected by the Earth's rotation, creating either northeasterly or southeasterly winds, called *trade winds*. Some of the warm air continues its flow toward the polar regions.

The size of these patterns might lead you to

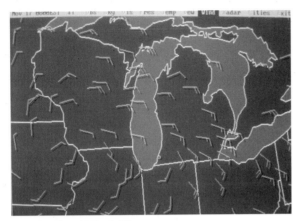
Surface Wind Chart

expect them to be highly stable, thus creating steady, predictable weather patterns across the globe—and they would, except for a few factors. First, the surface of the Earth is neither uniform nor smooth. Air heats differently over land and over oceans, so the large air masses in the doldrums that rise toward the polar regions do so at differing rates.

The Earth's spinning constantly changes the landmass under an air mass, and the air in it never rises at a steady rate. Also, various elements of the geography of the Earth—its mountains, valleys, and oceans—shape the wind flow differently. Finally, the Earth is tilted on its axis of rotation, thus causing seasons by the vast differences in the amount and location of heating that the sun brings to different regions. It's these differences that give the wind such fascinating qualities.

Fantastic stories have been told about the wind by societies throughout history. Most of the special qualities ascribed to wind live only in ancient folklore and have long since been proven to be no more than myth. Still, even in modern U.S. society, wind can earn a name or be said to waft enlightened guidance our way. The song "They Call the Wind Maria" is one example of giving the wind personality traits, as if its rushing air actually contained the breath of life. Peter, Paul, and Mary told us in the popular folk song "Blowin' in the Wind" that we can turn to the wind for guidance. But winds were wearing "name tags" long before our modern society existed.

The Chinook Winds

The Chinook Indians are native to the Pacific Northwest, primarily in the area that now holds the states of Washington and Oregon. European settlers first encountered them in 1792. Over the next 100 years or so, the Chinooks became skilled traders, expertly bartering with the increasing flood of explorers and settlers from the east. Their influence was considerable. Other Native American tribes as well as white explorers and settlers learned their language to facilitate commerce, and Chinook became the language of business.

This success soon proved to be the Chinooks' undoing. Late in the 1800s a series of epidemics, brought in by the new inhabitants, killed off most of the Chinook population. Still, their culture lives on through the well-known Chinook winds.

The *Chinook* is a warm, dry wind that blows down out of the Rocky Mountain slopes in winter and early spring. Since this wind comes down from Chinook country, the early settlers gave it the name of the natives of that land. The temperature of a Chinook wind rises rapidly as it roars down from the west onto the Plains, rising 1° for every 180-foot drop in elevation. Thus, the temperature of a typical Chinook that descends 5,500 feet will be about 30° F warmer than it was at the top of the mountain. Always welcomed by anyone living in the area, a strong Chinook can descend from the mountains, spread out at the mountain base, and raise temperatures enough to melt the snow and expose grass so that animals can graze.

When similar wind phenomena occur in other areas of the world, the winds are called *foehns*. Other ancient societies have hailed local winds as divine and have pegged their fortunes, their safety, and their futures on what the winds blew in. The following is a good example of a primitive belief about wind that lingered long after it had passed its useful life. An ancient story, it carried a larger-than-life message and had a dramatic impact on two of the most powerful nations in modern times: the United States and Japan.

CHAPTER 5: WIND

The Divine Wind

Great storms destroy, but they also can create. In the 13th century, two violent storms destroyed the armies of the Mongol emperor Kublai Khan and, at the same time, also created Japan's national myth of military invincibility that lasted down to the Second World War.

The First Invasion

Kublai Khan, grandson of Genghis Khan, ruled China from 1260 to 1294. Following a practice of earlier emperors, he demanded tribute from China's neighbors, starting with Korea and what is now Vietnam. Later—perhaps attracted by Marco Polo's fabulous accounts of gold, pearls, and other riches—he turned his attention to Japan. In a state letter to the ruling Hojo regents, Kublai demanded they pay tribute or face invasion. This letter, and several others that followed, went unanswered.

The letters, however, were not ignored. Nobles and military leaders took the threat of invasion seriously. They strengthened their defenses and offered up prayers at their temples and shrines.

Finally, Kublai Khan would wait no longer. In 1274, he sent a naval expedition to invade Japan. He sent a fleet of 900 vessels, carrying 40,000 troops—about eight times the force William the Conqueror sent to England. Some of the ships in use in the area at that time were of considerable size. According to the contemporary description by Marco Polo:

> Ships of the largest size require a crew of three hundred men; others, two hundred; and some one hundred fifty only, according to their greater or lesser hulk.

The expeditionary force landed on two small islands near Hakata Bay on the island of Kyushu. They terrorized the inhabitants, plundered whatever they found, then moved on to Kyushu itself. As the Mongol ships approached, every able-bodied Japanese man who had a horse and a sword was urged to hasten to Kyushu to meet the invaders.

The Mongol cavalry could maneuver skillfully in close formations and used poisoned arrows—tactics unknown to the Japanese. Mongol weaponry also included catapults that hurled heavy stones, possibly loaded with explosives. Altogether, their awesome military might threatened to overwhelm the defending Samurais, who were armed mainly with swords and courage. It appeared that the entire Japanese defending force might be annihilated before reinforcements could arrive.

When night fell, bad weather was approaching. The weatherwise Korean sailors accompanying the Mongols advised them to return to their ships and prepare for the next day's battle. But that battle was never fought. A fierce November gale, possibly a typhoon, suddenly struck and destroyed as many as 200 Mongol vessels, forcing the remainder to withdraw.

According to Korean records, cited by historian Sir George Sansom, 13,000 of the invading force lost their lives during the expedition. The Japanese defenders were saved by the storm—at least for the time being.

The Second Invasion

Anticipating another invasion, the Japanese began straightaway to construct a defensive stone wall around Hakata Bay and to build many small warships to counter the expected return of the Mongol fleet. Meanwhile, Kublai Khan spent several years subduing southern China before again turning his attention to Japan. First, he tried persuasion. Two Chinese envoys arrived in Japan and bade the Japanese rulers to appear in Beijing to pay homage to the imperial court. This arrogant demand infuriated the Hojo leaders, and the luckless envoys were summarily beheaded. Another envoy arrived soon after and suffered the same fate.

The executions made a second invasion inevitable. In 1281, the Khan sent two fleets, 4,400 vessels in all, carrying 140,000 men. This was the largest naval operation up to that time and probably is still one of the largest.

For an undeveloped agricultural society, the economic and environmental costs of this second Mongol invasion must have been staggering. Just to build all the ships, incredible amounts of timber had to be cut. A contemporary Chinese poet lamented that denuded hills all over China mourned for their lost forests.

The two fleets set out simultaneously in June 1281, one from Korea, the other from south China. The first fleet reached Kyushu in late June, and again the invaders landed near Hakata Bay. This time, however, the Japanese defenders were ready. They had completed their fortifications and now used their newly built warships to harass the invading fleet. Still, some of the invaders managed to land, and fierce fighting went on for several weeks. Finally, in early August, the second fleet arrived. The Khan's combined forces assembled at the mouth of the bay and prepared for a final assault.

Then it happened again! On the fifteenth of August, the sky darkened and a tremendous typhoon blew up. The great storm battered Kyushu with onshore winds for two days. The second fleet, operating in the Gulf of Imari, was exposed to the full fury of the winds. When, in a desperate attempt to escape, it made for open water, tidal surges jammed the helpless, massed vessels together and onto the shore. More than half the invaders were killed or drowned and their ships reduced to an appalling mass of splinters.

According to a contemporary Korean account, "The bodies of men and broken timbers of the vessels were heaped together in a solid mass so that a person could walk across from one point of land to another on the mass of wreckage." Other invaders, left behind on the shore, were slaughtered by the defending Japanese warriors.

Marco Polo described the second invasion. However, since he was a guest in Kublai Khan's court, his portrayal of the extent of the disaster is somewhat low-key, compared with Japanese accounts:

> It happened, after some time, that a north wind began to blow with great force, and the ships of the Mongols, which lay near the shore of the island, were driven foul of each other. It was then determined, in a council of the officers, that they ought to disengage themselves from the land; and accordingly, as soon as the troops were re-embarked, they set out to sea. The gale, however, increased to so violent a degree that a number of the vessels foundered. The people belonging to them saved themselves upon an island lying about four miles from the coast.

Emperors then assumed that their rule was a mandate from Heaven. Naturally, Kublai Khan was reluctant to believe that the mere forces of nature could overrule Heaven's supreme authority. Several years after the frustrated invasion attempts, on analyzing what had happened, the emperor convinced himself that dissension between his two commanders on the scene had led to their defeat. Accordingly, both commanders were executed. Whatever the cause, the result was that, following the death of Kublai Khan, the Mongols lost all taste for any further adventures in Japan.

The Divine Wind

These were the first attempted invasions of the Japanese islands in historic times and the last until the end of World War II. These events, especially the second storm, made a powerful impression on the people of Japan, who regarded their timely rescues as divine intervention to protect them from foreign enemies. The stories were incorporated into a patriotic national epic and the providential storms received the name "kamikaze," or "divine wind."

Passed down through succeeding generations, the epic taught that, in times of struggle, the Japanese people would always be able to overcome foreign enemies, no matter what the odds. Some 600 years later, in 1944, the persistence of this belief was brought home to American servicemen in the Pacific when Japanese "kamikaze" pilots dove their explosive-laden planes into the ships of the U.S. Navy.

Typhoon Roulette

Given the weather patterns in the area, how daring were the Mongol invasions? How might a maritime insurance company today respond to a request to insure an operation, even in peacetime, that involved 4,400 heavily loaded, motorless vessels crossing the East China Sea at the height of the typhoon season, without access to weather data or communications?

According to modern meteorological records, the region of the northwest Pacific that borders Japan and the south China coast is the most prolific generator of large cyclonic storms in the world. Of all such storms worldwide, on the average, 37 percent occur in the northwest Pacific–Asian border region. As a comparison, hurricanes, the corresponding storms in the Caribbean region, make up only 12 percent of the total. During the period 1949–1976, over the entire northwest Pacific, there were, on the average, 29 typhoons per year. Many of these storms eventually reached Japan or the China coast.

Broken down by months, the average is 6.1 typhoons in August and 2.8 in November, the months in which the emperor's fleets came to grief. The invading forces were either at sea or fighting near the shore for periods ranging from several weeks to several months, during the season of greatest typhoon risk. From this perspective, the invasions would appear to have been very hazardous operations. And why try again after the first fleet was lost? Following their improbable conquest of most of Asia on horseback—ranging from the Pacific coast westward all the way to Poland and Hungary—the Mongols also had a patriotic epic of invincibility. And when two "invincible" forces collide, Nature often decides the outcome.

—By Richard Williams. Reprinted with permission from *Weatherwise*.

World Record Winds

- *Fastest surface wind speed recorded:* 231 mph
  Place: Mount Washington, New Hampshire
  Date: April 12, 1934
- *Windiest place:* gale winds regularly exceed 200 mph
  Place: Commonwealth Bay, Antarctica
- *Fastest tornado winds:* 286 mph
  Place: Wichita Falls, Texas
  Date: April 2, 1958
- *World's worst tornado:* killed 792
  Place: South-central United States
  Date: March 18, 1925
- *Fastest hurricane winds near a storm center:* 74 mph
- *Hurricane with the highest wind gusts:* 175–180 mph

> *Place:* Central Keys and lower southwest Florida coast
> *Date:* August 29–September 13, 1960
> - *World's worst cyclone:* unleashed floods that killed 200,000
> *Place:* Bangladesh
> *Date:* 1970

Windblown Verses

Many poems and songs sprang out of the ancient mariners' times at sea and their total vulnerability at the mercy of the weather. While the term "horse latitudes" has the ring of a tall tale, its deadly toll on so many horses makes it all too real (see Chapter 9, p. 121). Yet not all old sailors' yarns about the weather were so gruesome. Some were accurate and useful lore that perhaps saved many lives over the centuries. Here are a few examples:

- A northern air brings weather fair.
- An honest man and a northwest wind generally go to sleep together.
- The west wind is a gentleman and goes to bed.
- A veering wind will clear the sky,
  A backing wind says storms are nigh.
- When the wind is in the north,
  The skillful fisher goes not forth;
  When the wind is in the south,
  It blows the flies in the fish's mouth;
  When the wind is in the east,
  Then the fishing is the least.
  But when the wind is in the west,
  There it is the very best.

For centuries observing the wind and how it shifts has allowed reliable weather predicting. When high- and low-pressure systems approach, seamen have long noticed changes in wind direction.

With a high-pressure system, winds circulate clockwise and shift from the southwest to west to northwest, setting up a *veering wind*. This is a sign that fair weather is ahead. Wind direction is the opposite for a low, moving counterclockwise from northeast to north to northwest. This is called a *backing wind* and signals the approach of inclement weather.

Local weather also can be revealed by wind direction. In the Northern Hemisphere, for example, an east wind often brings nasty weather. Not all wind effects, however, can be predicted by catchy sayings. Sometimes the wind does need to be the subject of ancient folklore to be memorable. In its most violent form, no one needs a poem to aid recall; some things in nature are beyond forgetting, and tornadoes are at the top of the list.

Tornadoes: Heavyweight Prizefighters

Hurricanes are larger and cause more extensive damage overall. Blizzards last longer and can bring life to a standstill over vast areas. But nothing in nature inspires as much sheer, acute terror as a tornado.

While not everyone sees a tornado in person, those who do never forget the moment they saw their first funnel cloud buzz-sawing its way through anything in its path. Here's a dramatic account of a recent, classic tornado that defined "opening day" for tornado season in Tornado Alley in 1991.

Countdown to Disaster: Chasing the Andover Tornado

April 26, 1991, dawned cloudy and windy. It looked like a dreary, rainy day in Dallas, Texas. The window screens were rattling on the south side of the house and power lines howled in the stiff breeze.

As I do on most chase mornings, I tuned to the A.M. Weather program on public television. A major storm system was forecast to strike the high plains. It was a classic severe weather situation, with low pressure in eastern Colorado and a dryline extending southward into West Texas. I was certain Mother Nature had set the stage for the first tornado outbreak of the season.

I already had checked my chase equipment. Cameras had been cleaned, film purchased, and batteries recharged. Other chasers had called to compare notes. My adrenalin was pumping and excitement was in the air. The only question was where would the storms form?

The National Weather Service was forecasting a high risk of severe thunderstorms and tornadoes over much of the central and southern plains by afternoon. I began plotting hourly surface weather observations across Kansas, Oklahoma, and Texas. The best chance for tornadoes appeared to be in southern Kansas, a seven-hour drive away.

All the weather features seemed favorable for severe weather. The jet stream was overhead. Cold temperatures aloft would ensure an unstable atmosphere. The winds spiraled and increased with height (proper ingredients for rotating storms). And low stratocumulus clouds were racing northward across the Dallas skyline, carrying the moisture that would fuel thunderstorm development later in the day.

The Chase Begins

I called Carson Eads, my chase partner. Carson is a ham radio operator and has a custom "chasemobile" complete with two-meter and high frequency radio transceivers, a portable weather station, and color television. A trailer towed behind the vehicle carries a tapered metal stand supporting a rotating, telescopic antenna. We call it the "Eiffel Tower." The antenna boosts reception so much it allows Carson to talk to people in the next state!

Carson and other chasers arrived at my house around 10 A.M., and we immediately headed north on I–35. We heard the first severe weather broadcast of the day as we crossed into Oklahoma:

A potentially dangerous severe weather situation is developing . . . as a powerful storm system moves into the plains states. Moist and very unstable air is moving rapidly northward into Oklahoma and Kansas . . . . The dryline will begin moving eastward by about midday as strong upper-level winds over 100 mph swing around the south side of the storm system and into the southern and central plains. By early afternoon, thunderstorms are expected to develop rapidly along and ahead of the dryline across western Oklahoma and spread into central Oklahoma by late afternoon and early evening. . . . Extreme instability and strong winds aloft indicate the potential for a significant severe weather outbreak later today including the possibility of very destructive tornadoes . . ."

We arrived in Norman, Oklahoma, around 1 P.M., pausing briefly to top off the gas tank and grab some groceries. The latest satellite image showed a cluster of thunderstorms developing in northwest Oklahoma west of Enid. Within minutes, the National Weather Service issued a tornado watch for a large portion of Oklahoma and Kansas. I was concerned about the damping effect of the high-level cirrus cloud cover overhead and believed we needed to continue north into Kansas, where clear skies prevailed.

Tail-End Charlie

As we crossed the Kansas border around 4 P.M., we sighted the first storms developing to the west. However, the storm tops were being ripped apart by strong wind shear. The radar display on the television showed a line of thunderstorm cells extending southwestward to the Oklahoma border. A severe thunderstorm warning had been issued for the storm at the southern end of the line, a position chasers call "tail-end Charlie."

We left the Interstate south of Wichita and headed west toward tail-end Charlie. The sky was milky white, veiled by a thick haze bleached by the sun. At 5 P.M., the severe thunderstorm warning was reissued for the county just to our west. As we approached Argonia, Kansas, the sky darkened and a rain-free cloud base came into view just ahead—a ragged, turbulent cloud base with a tail cloud extending off to the north. In an instant, I knew it was the classic tornado storm structure. White-hot lightning bolts zagged through the darkening blue sky in increasing tempo. "No question we could have something here," I said. We pulled off the road and grabbed our video cameras. There was a stiff east wind, and pea- to marble-sized hail was falling. I focused my attention on the wrapping rain curtains to our southwest. Black clouds were boiling in what seemed to be time-lapse motion. "We are going to have a tornado here!" I shouted. Sure enough, at 5:15 P.M., an elephant-trunk-shaped funnel dipped towards the ground. Immediately, the sirens sounded in nearby Argonia. My heart rate quickened and my breathing became shallow—the tornado was heading right at us! Fortunately, it was several miles away, but we kept the motor running just in case.

In picture-perfect contrast, we began filming a black tornado against a white cloud background. For 10 minutes, we watched in awe before the tornado shriveled and dissipated.

Entering the Bear's Cage

The storm was heading northeast, on a direct course to Wichita. Local television stations were still warning of the tornado we had just witnessed. Radio scanners were buzzing with spotter reports about a new cloud "lowering" gathering over the town of Conway Springs.

There were no direct roads, so we lost valuable time by having to first drive east, then north. In the process, the storm beat us to the town. To keep pace with the storm Carson and I would have to enter "the bear's cage," the wrapping curtain of rain on the back edge of the storm that could hide a tornado.

My hands grew cold as we drove into the rain. The visibility dropped dramatically and strong north winds buffeted our vehicle. We turned east and in a few minutes the rain ended as we passed through the town of Clearwater. A large cloud lowering had massed on the east side of town. We came upon an open field where all the clouds came together. Suddenly condensation shot upward from the ground about a mile to our southeast. "Multi-vortex tornado!" I shouted. Spotters immediately relayed their reports to the local National Weather Service office.

The tornado crossed the road in front of us and hit a house. The roof disintegrated and a plume of attic insulation was sucked into the vortex, appearing like smoke from a fire. In anger and disbelief, I witnessed the destruction of two more farmhouses. "Damn, another house just went down there," I muttered. Never had I felt so helpless. And other houses still lay in the storm's path.

Carving out a swath of destruction, the tornado ravaged homes in the northwest part of Haysville, then entered south Wichita. In seconds, homes disappeared from their foundations. Broken plumbing lines created water geysers where homes had once stood.

The tornado then turned east, striking McConnell Air Force Base and just missing rows of parked fighter planes. We tried to keep pace with the twister to no avail; it had toppled power poles across the road in front of us, cutting off our pursuit.

The town of Andover was next on its hit list. We could hear spotters frantically telling emergency officials to look to the southwest. A policeman in the town responded, confirming that a large tornado was bearing down on them. With no operational sirens, one officer tried in vain to warn residents of the Golden Spur Mobile Home Park on the edge of town to evacuate. Unfortunately time ran out for many of them at 6:35 P.M., when the tornado obliterated hundreds of homes and left scores of dead and injured.

—By Tim Marshall. Reprinted with permission from *Weatherwise*.

Major Tornadoes

The ancient Roman adage to "beware the Ides of March" is good advice today for everyone living in tornado territory. The Ides is officially March 15, and it marks the beginning of a short span on the calendar that's spawned the most death-dealing tornadoes on record.

The deadliest series of tornadoes on record occurred during the late afternoon on March 18, 1925, in portions of Missouri, Indiana, Illinois, Kentucky, and Tennessee. Eight separate tornadoes were observed. One of those killed 689 persons, injured 1,890, and caused more than $16 million in property damage (this was a time when new cars could be bought for a few hundred dollars and a home cost but a fraction of what a car costs today). The other seven tornadoes of the series increased the total loss of life to 740 and contributed significantly to the total casualty and property damage.

Another major series of tornadoes killed 268 persons and injured 1,874 in Alabama on March 21, 1932. Property damage amounted to approximately $5 million.

More recently, in 1990, March storms spawned a series of four separate tornadoes in central Kansas. Two of these tornadoes joined forces at one point and resulted in one of the three most intense tornadoes in recent history. This monster series of intertwined tornadoes cut a swath more than 100 miles long as it chewed up the ground during its two-and-one-half hour reign of terror.

The Great Earth Mover

Even with all our mighty steel foundries, humans will never be able to build a machine that can move earth as powerfully as the invisible wind. The wind is nature's "Earth mover" and helps to rearrange the planet completely every few million years or so.

If you have ever felt a strong wind whipping small pieces of debris through the air, pelting away at your exposed skin and causing a smarting pain, it will be easy to understand how wind attacks the land surface to modify the landscape.

Wind alters the landscape in the desert by pounding away at boulders, slowly crushing them into stones, pebbles, sand, and dust. Then the wind blows these particles away, flinging them at whatever is in the way and dropping them elsewhere.

Weathering causes stones to break away from the larger rocks in the desert. This gravel is then moved when the wind blows and blasts against the other rocks, causing the stones to break into sand grains. The wind does not lift sand very high—only about to waist level—but these sandstorms are so powerful that they wear away the rock from the bottom, creating what are known as *mushroom rocks* (because the tops are bigger). *Yardangs*—unusual stone desert corridors—are formed when the blowing sand digs long, narrow trenches in the soft rock layers, leaving the hard rock layers standing vertical.

The desert hills are not worn down to flat levels of rock, but the fierce wind lifts up and carries away the rock, leaving deep holes. One such "hole" in the Sahara Desert, the Qattara Depression, is quite large (it rivals the size of Wales), and so deep that parts of it are below sea level.

Eventually, the sand that is being transported by the wind is dropped when something blocks its path or when the wind dies down. The landscape is built up into one of several types of sand dunes when a continuous wind blows from one direction.

Barkhans are dunes that resemble the shape of a new moon, because blowing wind has moved the sides forward, forming pointed horns. It is not unusual for a barkhan to reach several stories high and for it to creep across the desert at up to 50 feet per year.

Barkhans sometimes become rows of long dunes when the wind shifts. The desert peoples call them *seif* (meaning "sword") *dunes* because they remind them of swords with wavy blades. In Iran and Algeria some of these dunes are higher than a typical block of skyscrapers in a big city.

In the desert there also are areas where smooth sand covers the land. This happened when pebbles caused the wind to drop the sand, thereby spreading it out.

Rocks that are pummeled by wind create dust. The wind carries dust particles farther than it can carry grains of sand. The windblown dust layers that move across the Earth are very thick and cover much of northern China, the middle of the United States, and Europe between France and Russia.

The wind also indirectly helps redesign our landscape by carrying airborne seeds. Plants spread their offspring across vast distances with the help of the wind. When the offspring plants take root in new areas, they, too, can change the landscape as their roots penetrate into crevices of large rocks, splitting them apart. As the larger rocks break down, some pieces eventually become small enough that the wind can move them. It's a never-ending process that carves out enormous works of art and levels mountains—dust particle by dust particle.

Whirlwind

Tornadoes, also called cyclones and whirlwinds, have inspired many exciting Native American tales. One story, originating with the Blackfoot Indians but shared by many tribes, tells of an orphan boy who rescues people who have been devoured by Windsucker. Windsucker was a whirlwind that was said to be a giant sucker fish. The boy kills Windsucker by doing a jumping-up dance inside the monster's belly with a knife, fastened blade-upright, on his head.

The Seneca named the tornado *Dagwa*

CHAPTER 5: WIND

Noenyent and said it was a giant rolling head that could tear the largest trees from the earth.

The Kiowa Legend

The ancient Kiowa Indians claim that their people are the ones who made the first tornado and that if they spoke to it, the tornado would pass without harming them. Here is their whirlwind legend:

Hot summer lay upon the people like a heavy moist blanket. Not a leaf moved. No whisper of a breeze lifted a strand of hair.

The people gathered around an old medicine man. "Tell us what we can do. There is hardly a breath in this hot, stuffy air," they said.

The old man sent some women to the river bank to gather red mud. When they returned, he told the people to watch carefully. Then he shaped the mud into the form of an animal. It looked something like a horse, with four legs and a long tail.

"Watch me, and do what I do," said the old man to the people who were gathered around. Then the old man blew into the nostrils of the horse and it began to grow.

"Blow hard!" the old man commanded the people. They blew at the horse as hard as they could, and it grew larger still until three people were needed to hold it, then ten. As they held the red mud horse, it began to stretch and twist.

"Red horse, I name you Red Wind!" called out the old man. "Now show us what you can do to cool the people."

With that, the horse tore away from the many hands that were holding it and whirled through the air, stirring it and laughing and blowing up dust.

At first, the people cheered and lifted their arms to the cooling gusts. But the wind blew the feathers from their hair and the bracelets from their wrists. Red Wind twisted and jumped like a giant wild stallion. Trees snapped as though they were twigs.

"Now look what you've done, old man!" shouted the people. "Red Wind will tear up the earth and blow us all away!"

So the old man called out again, "Red Wind! I made you. I named you. Now I give you a home. From this time on, you will live in the sky."

Ever since, Red Wind, the whirlwind, has lived among the black clouds. But sometimes he revisits his old home land and twists and stretches down because he likes to kick up things on earth.

Wind-Related Products

W-10 Wrist Anemometer
World Market
P.O. Box 19941
Cincinnati, OH 45219

Now you can determine wind direction and speed at any time. Just carry the W-10 and you can instantly read out wind information using this stopwatch-size instrument. It's perfect for storm chasers and useful for anyone who wants to keep up to date on the weather. The W-10 is calibrated in knots, miles per hour, meters per second, and the Beaufort Scale. It includes a neck lanyard and is priced at $24.95 plus $1.50 shipping and handling.

Sou'Wester Anemometer
American Weather Enterprises
P.O. Box 1383
Media, PA 19063
(215) 565–1232

This one won't fit on your wrist, but it will mount easily on a rooftop or mast. For outdoors it includes a three-cup rotor sensor, mounting hardware, and 60 feet of wire. For indoors it includes a wood case with mahogany finish with hardware for shelf or wall mounting. The speed scale reads from 0 to 105 mph and from 0 to 90 knots. It's made in the United States by Maximum, comes with a one-year warranty, and is priced at $119.95, including shipping.

Copper Reindeer Weather Vane
The Nature Company
Catalog Division
P.O. Box 188
Florence, KY 41022
(800) 227–1114

Traditionally styled homes can seem so much more traditional with a weather vane on top. The Nature Company offers a hammered-copper reindeer weather vane that mounts easily atop a roof, fence, or post. Not only will this vane, which costs $148, provide minute-by-minute readings on the changing wind, it will also add an authentic touch of the past.

Exotic, Resonant Wind Chimes
The Nature Company
Catalog Division
P.O. Box 188
Florence, KY 41022
(800) 227–1114

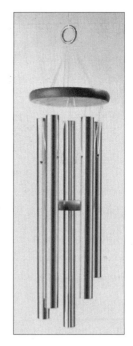

Of course wind chimes aren't precision weather instruments, but they can bring you a lot of enjoyment by giving a voice to the wind. Created by Woodstock Percussion, these exotic chimes respond to the slightest breeze to produce rich melodies from a variety of exotic musical scales.

They have the 15-inch "Blue Note" chime, $29, that is based on the traditional American blues scale. The 24-inch "Bali" chimes, $39.95, use scales from the Balinese shadow play.

The 36-inch chimes of "Olympus," $65, are tuned to the pentatonic scale. The stately granddaddy of their selection, the 55-inch "Westminster" chimes, $76, create velvety peals of classic tones.

Weather Vane Plan Books
Wind & Weather
The Albion Street Water Tower
P.O. Box 2320
Mendocino, CA 95460
(800) 922–9463

Whirligigs & Weathervanes, by David Schoonmaker and Bruce Woods, Sterling Publishing, 1991. Make your own, unique weather vanes and wind toys

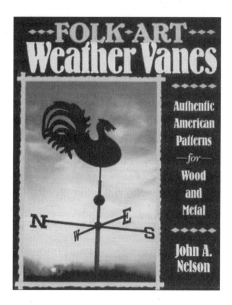

with this book, which contains easy-to-follow patterns for more than twenty whimsical wind gadgets for you to create. This book contains 128 pages, is hardbound, and has a large format. The price is $19.95.

Folk Art Weather Vanes, by John A. Nelson, Stackpole Books, 1990. Create your own, inexpensive copies of traditional weather vanes with this book, which contains detailed patterns for making sixty-eight complete weather vanes. Each pattern is modeled after an original American folk-art masterpiece. This book contains 159 pages, is softbound, and has a large format. The price is $16.95.

Nimbus Wind Monitor
Sensor Instruments Company, Inc.
41 Terrill Park Drive
Concord, NH 03301
(800) 633–1033
In New Hampshire: (603) 224–0167

Sensor has recently updated its Nimbus Wind Monitor, giving it more accuracy, more computations, and more memory features than ever before. It combines microcomputer technology with the proven reliability of lexan anemometer cups and reed switch directional vanes. Electronic filtering evens out low-speed fluctuations yet provides rapid response to gusts. Its memory will store average speed, prevailing direction, and peak gusts (with direction) for every hour over a ten-day period and the peak gust over the last five minutes and fifteen minutes.

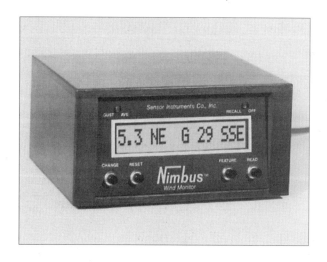

Housed in a cherry or oak case, it is AC powered, with a six-week battery backup. Speed range is 0–150 mph and can be read in knots, kilometers per hour, meters per second, or feet per second. The direction indicator uses a sixteen-point compass. It comes with a three-year warranty and has an optional computer interface. The monitor alone is $535; with the sensor kit it is $735.

Wind Speed and Direction Recorder
Robert E. White Instruments, Inc.
34 Commercial Wharf
Boston, MA 02110
(800) 992–3045

For wind-energy studies, small airports, pollution monitoring, ballooning, smokestack industries, flying clubs, and yacht clubs, the Controlex #A11-131 from White records wind speed and direction on a continuous, moving paper chart. Using two pens on one chart, this recorder graphically displays information full-time, over a thirty-day period. Speed range is selectable between 0–50 mph and 0–100 mph. The unit comes in a rugged, metal carrying case with a handle and is powered by 110-volt AC, but a truly portable DC-powered version is available. The total package includes remote sensors and 100 feet of connecting cable. This is a professional unit; call White for current pricing.

Davis Weather Wizard III
Davis Instruments
3465 Diablo Avenue
Hayward, CA 94545
(510) 732–9229

The Weather Wizard III monitors indoor and outdoor weather conditions and displays them at the touch of a button. Outside sensor pod with a stand-off arm is included with mounting hardware. And it includes a 40-foot sensor cable for remote placement, but with an optional cable you can place the sensors up to 160 feet away. It includes these important weather-watcher features:
- All highs and lows are recorded with the date and time.
- Selected weather readings can be scanned automatically.
- Choice of Metric or English units of measure.
- Can be mounted on a desk, shelf or wall.
- Uses a 9-volt battery backup to retain recorded entries even during power outages.
- Can be connected to a personal computer using the Weatherlink software for data storage, analysis, and graphing.

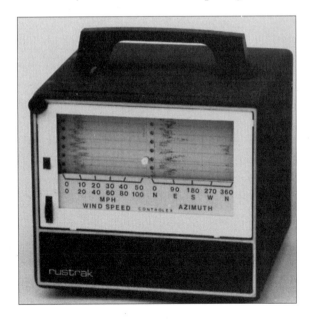

Chapter 6
Temperature

Temperature is the fuel gauge of our atmosphere. It is our way of measuring the energy that the Earth has absorbed from the sun. Just as with a car's gas tank, more "fuel" in the atmosphere's "tank" means more energy available for creating weather.

Or you might consider temperature measurements analogous to measuring how tightly the spring has been wound on a windup toy. The tighter it's wound, the more the toy will do. The enormous energy that the atmosphere absorbs from the blazing overhead sun in hot climates during the summer is like an overwound spring. Once it is unleashed you never know for sure what will happen.

In this chapter we'll look at how temperature changes, get a special course on how to measure and record it accurately, and learn one weird way to measure it not so accurately. We also will learn what temperature does to human physiology as well as take a look at how it affects global weather patterns. Finally, we'll list some all-time record high and low temperatures and then describe some temperature-sensing devices that go beyond being mere thermometers.

How Temperature Changes

Solar radiation is the single greatest cause of temperature variations. We all are familiar with the increase in temperature during daylight hours and the decrease with nightfall. Many other factors, however, also have a major influence on temperatures. These include the heating of land and water, ocean currents, altitude, and geographic position. Let's look at each one briefly.

Land and Water

Examining the heating properties of various surfaces helps us to understand variations in air temperatures, because the heating of the Earth's surface controls the heating of the air above. Land heats to higher temperatures more quickly and cools to lower temperatures faster than water. Therefore, air temperatures vary more over land than over water.

Water is highly mobile, so surface water temperatures rise and fall more slowly. The heat is distributed by water turbulence. Temperature changes occur daily down to nearly 20 feet below the surface and yearly down to depths of up to 1,800 feet in oceans and deep lakes.

On the other hand, heat does not penetrate deeply into soil or rock; even large, annual temperature variations usually reach to depths of less than 50 feet. But this relatively thin top layer on land is heated to much higher temperatures in the summer and during winter cools more rapidly than water.

Transparent water allows solar radiation to penetrate more deeply, and water requires more heat to raise its temperature the same amount as an equal piece of land. Evaporation—which causes cooling—is greater from water than land. Water warms more slowly, stores greater quantities of heat energy, and cools more slowly. All these factors add up to differences in air temperature over a large body of water, compared with land under the same air mass.

Ocean Currents

The effects of ocean currents on temperatures of nearby land areas vary. It is well known that poleward-moving warm ocean currents have a moderating effect. Winter temperatures in Great Britain and much of western Europe are warmer than might be expected for their latitude, due to the so-called North Atlantic Drift. The moderating effects are carried inland by the prevailing west winds, thus increasing temperatures over the landmass.

The effect of cold ocean currents on landmass temperatures is most noticeable in the tropics and, during the summer months, in the middle latitudes as compared to the effects of warm ocean currents, which have their largest impact on landmass temperatures during winter.

Altitude

Changes in altitude bring about a great variation in temperature. Temperatures drop at higher altitudes, and atmospheric pressure and density also decrease.

From sea level up to the highest altitudes in which humans live, temperatures drop at a fairly constant rate. This steady decrease in temperature is called the *adiabatic lapse rate*. Temperatures decrease at the rate of 3.5° F per 1,000 feet of altitude, or about 1° for every 300 feet. (The lapse rate is 2° C per 1,000 feet.)

The thinner air of higher altitudes is less dense than at sea level. Thus, it contains fewer molecules to heat. This thinner air leads to more rapid and intense daytime heating, followed by more rapid cooling at night in high mountain locations than at lower altitudes.

Geography

The effects of geography on weather are as varied and detailed as geography itself. Mountains act as barriers and can divert, destroy, or help form storms and can even shift the high-flying jet streams. (Jet streams are strong westerly winds that flow in relatively narrow and shallow streams at high altitudes.)

Mountains cause air to rise as winds strike them, and the winds are driven up their "ramps" into the sky. They also can trap air in one location and keep the air over a region from changing. Los Angeles, Calif., is an excellent example of this phenomenon. Ringed with mountains, the L.A. air can "park" for long periods and fill with pollutants instead of being swept clean by ocean breezes.

The temperatures experienced at a specific location are greatly influenced by the geographic setting. For example, one location on a coast that experiences wind blowing in from the ocean will have quite different temperatures from a nearby coastal location where the wind blows from the land to the sea.

Ice Man

Ice Man, Cold-Bringer, and Winter Man were all names that Native Americans gave to the fierce spirit of the cold months. In a Sanpoil legend, Northern Lights was the ruler of the cold. He had five cold sons, who each went out in their turn to bring the winter kill farther and farther south.

There was coal in the mountains of the Cherokee Indians. Their winter tale seems to be about a fire that burned deep into a vein of coal that lay on their lands.

The Cherokee farmed in the southern Appalachian Mountains until many were forced to sell their lands and move west to the area we know as Oklahoma. But others remain in their mountain homeland even to this day. Here is their Ice Man tale:

CHAPTER 6: TEMPERATURE

One fall, the people started a big fire to burn off the underbrush. A tall poplar tree caught fire and burned long after the people had gone. It burned the branches and the trunk, and when there was nothing left, the fire settle into the roots and burned a hold in the ground.

The next day men passed pots of water from the river to the hole. The women threw in dirt and rocks. But still the fire burned until the small hole in the ground became a huge hole.

The people peered deep in the hold, asking fearfully, "Will this fire eat up the whole world?"

Then one of the elders spoke. "Far north lives fierce Ice Man, who sends our frozen season. Maybe cold will kill the fire." So some of the people marched the long way north to ask for Ice Man's help. Creeping cautiously to his ice house, they found a small man with white hair hanging to the ground.

"Can this little man help us with such a big fire?" they wondered.

"Yes, yes! I can help you," he exclaimed, quickly unbraiding his long hair. The he took his hair in one hand and whipped it across the other. At once a cold wind began to blow. He struck his hand with his hair a second time, and a light rain began to fall.

"A little rain won't kill that fire," cried the people. So Ice Man shook his hair fiercely and raindrops turned to sleet. With another shake, hailstones pelted the ground, bouncing around the people's feet.

"Go back now," Ice Man commanded. "When you are home, I will take care of your fire."

So the people returned home and found the others still watching helplessly.

The next morning as everyone stood before the glowing pit, a cold wind swept down from the north. The people shivered and shook, but the wind only made the flames blaze higher. Then a light rain began to fall. But the fire laughed and hissed and threw up clouds of smoky steam.

Now Ice Man was angry. He mixed sleet with the heavy rain and shook down snow. But the fire hissed louder. When the flames had eaten up the snow, Ice Man whipped his hair and threw cold winds at the flames until hail covered the fire and filled the hole. The people shivered and cried in their lodges as the wind became a whirlwind, driving rain and sleet into the ground and finally killing the last of the burning embers.

When it was over, the people crept back to the place where fire had been burning up the world and found instead a deep and beautiful lake.

Ice Man blows down from the north from time to time to make sure greedy Fire was really killed. But some of the people say that on quiet days, from deep in the lake, comes a faint sound of embers still crackling.

Temperature Measurement

The basic liquid-in-glass thermometers for measuring surface temperature are the mercurial maximum thermometer, the alcohol-filled minimum thermometer, the sling psychrometer consisting of two matched thermometers mounted on a common frame that is then rapidly whirled to ventilate the thermometers adequately. These are being augmented or replaced by electronic instruments that are quite reliable.

The familiar "U-tube" maximum and minimum thermometer (Six's thermometer) also is available but it can have an unstable error because it uses alcohol as the actual medium that reacts to temperature changes and propels a thread of mercury that propels the indices that indicate, in turn, the temperature extremes. The alcohol "wets" the sides of the tube causing, at times, a varying temperature error.

Although it is too unreliable for official use in National Weather Service programs, if properly designed, it can provide fairly reliable service adequate for amateur programs with a slightly wide margin of error.

Procure thermometers made to National Weather Service specifications. Although relatively expensive, they are accurate to within plus or minus 1° F. For obtaining basic surface temperature data you can rely on their indications if they are properly exposed. It is important to look for separations in the mercury or alcohol columns which can occur as a result of roughness of handling in shipping. The column can usually be reunited by striking the back of the thermometer frame repeatedly against the fleshy part of the palm of your hand until the column is reunited. This process will require patience, especially in the case of alcohol thermometers.

In regard to alcohol thermometers, it occasionally happens that part of the column will evaporate and then condense into liquid at the top of the tube. The treatment is the same for this condition, although more patience may be required. Always check for this condition or for column separation when resetting the extreme thermometers; they should agree after resetting to within 1° F.

Simple Thermometer Tests

Occasionally a NWS pattern maximum thermometer may become a "retreater," i.e., it may not retain the maximum temperature it has reached. Check for this condition occasionally by observing whether or not its column retreats when the thermometer is held in a vertical position after having been artificially heated to a temperature above the ambient temperature. If it does, it must be replaced. The column of a thermometer which is separated can sometimes be reunited more quickly by securing the thermometer to a stout cord or chain and whirling the instrument rapidly. The resultant centrifugal force will almost always reunite the column.

You may want to ascertain the error, if any, of the freezing point of your thermometer. This is something which it would be desirable to do on an annual basis, as some thermometers develop a gradually increasing error at their freezing point because of imperfect annealing of their glass tubes.

To do this, you need a sufficient quantity of frozen distilled water (not from the household tap), a styrofoam container of sufficient size to hold the amount of melting ice-and-water mixture and the thermometer to be tested (the thermometer should be immersed in the mixture up to the freezing point on the scale). The mixture should be stirred or agitated until the thermometer reading stabilizes. The difference, if any, from the freezing temperature of 32° F or 0° C should be recorded to the nearest 0.1° F or C. This method can be used also in determining the freezing-point error of electronic thermometers or other thermometer using a remote sensor; the sensor is submerged in the agitated melting ice and water mixture until the indicator stabilizes.

The Six's thermometer can also be tested in this manner, but it is important to submerge the instrument beneath the agitated mixture. The National Weather Service specification maximum thermometer can be subjected to this test also, but must first be done by placing the instrument in the freezing section of your refrigerator where the thermometer will be chilled to near 0° F. Quickly shake the thermometer down to a point below 32° F or 9° C and plunge it into the ice/water mixture until the reading stabilizes. Record the error, if any, to the nearest 0.1° F or C.

Determining Dew Point

The thermometers used in the National Weather Service sling psychrometer are made to the same specifications as are the standard NWS maximum and minimum thermometers. At high temperatures this perhaps is a sufficient standard, although it would be desirable even at those temperatures to have determined the errors of each thermometer at 10° or 20° intervals. However, at temperatures below freezing it is essential in the name of accuracy of dew point and relative humidity

CHAPTER 6: TEMPERATURE

determinations. The wet and dry bulb thermometers are read to the nearest 0.1° F or C for computational purposes. At low temperatures, very small differences in the readings between the wet and dry bulb thermometers can indicate significantly large intervals of dew point and relative humidity.

The determination of corrections at 10° or 20° intervals is generally beyond the scope of the amateur, since it requires precision standard thermometers and control equipment to maintain the temperature of comparison baths at very close tolerances. The National Bureau of Standards of Washington, D.C., will do this determination for a fee. Or you may ask the manufacturer of your thermometer who may make the determination for a fee. Otherwise, presume a 0.0° correction or use the correction determined for the freezing point as described above.

If you have autographic equipment such as a thermograph or hygrothermograph, it is important to set the instrument to zero correction at the beginning of a chart by setting the pen(s) to agree with the current temperature and relative humidity. At least once a day at a given time make a comparison between the indication(s) of the thermograph or hygrothermograph and the temperature and relative humidity as determined by the sling psychrometer. Make a time check on the record chart by depressing or raising the thermograph pen about an eighth of an inch. The hygrothermograph pen should NOT be depressed to avoid stressing the hair bundle which constitutes the sensor of such instruments; hygrograph pens should be RAISED to make the time check. Making such time checks affords a means of determining timing errors and correcting therefore when abstracting data from the chart. The appropriate instrumental corrections should also be placed over each of the time checks.

When determining dew point and relative humidity, keep in mind that these elements always are determined with respect to water, never with respect to ice. In the old tables WB No. 235 entitled "Tables for the Determination of the Relative Humidity, Vapor Pressure and Temperature of the Dew Point," Government Printing Office, Washington, D.C., the dew point values for 32° F and below are according to ice as are relative humidities at dry bulb temperatures of 32° F and below.

Apply a correction to these values to adjust them to vapor pressures according to water. Determine this correction using Smithsonian Tables 100 and 101 for converting between relative humidities with respect to ice and with respect to water and Table 102 for similarly converting dew points. An easier method is to use a psychrometric calculator—a circular slide rule—that directly reduces all psychrometric data to dew points and relative humidities according to water. These calculators are expensive (about $100) and may be procured from a reputable instrument supply house such as Qualimetrics, Inc. (call Novalynx Corp. (916) 477-5226, 1165 National Drive, Sacramento, CA 95834); but they are great time-savers.

It is also important to change the muslin sleeve on the wet-bulb thermometer regularly, as it gradually becomes contaminated with mineral deposits, especially in a marine environment, which changes the rate of evaporation of the water on the wet bulb and, consequently, the wet-bulb reading used in the computation of the dew point and the relative humidity.

Likewise, the instrument shelter should be painted periodically with a semi-gloss white paint inside and out. The National Weather Service suggests an annual painting. Also, the inside of the shelter and the instruments therein should be dusted periodically and kept clean.

—By Rev. Robert Duane and Tom Johnson.
Reprinted with permission from the *American Weather Observer*.

How NOT to Measure Temperature Accurately

The humiture factor certainly can illustrate how temperature affects humans. The hotter it gets, the slower we get. This is not true of all living creatures, though. Some act as though they love the heat and even sing its praises faster and faster as the temperature rises.

What creature possibly could love heat enough to sing about it? Think of a summer night. Close your eyes. What do you hear? Crickets. You can count on them.

Not only are crickets ubiquitous during the summer, you also can count on them enough to use their chirps to calculate the temperature in degrees Fahrenheit. All you need is a watch with a second hand and the following formula, where T is the temperature and N is the number of cricket chirps in one minute.

$$T = \frac{50 + N - 40}{4}$$

To get temperature in degrees Centigrade, use this formula instead:

$$T = \frac{5(N + 8)}{9}$$

This technique has been used for generations, but leave it to today's high-tech society to find a flaw with the basic premise. It turns out that you cannot randomly use any old cricket as an accurate substitute for a thermometer. Back in 1897, when A. E. Dolbear, a college physics professor in West Virginia, wrote the original formula, he based it on the chirping of the snowy tree cricket. Other species may not produce results with the same accuracy.

Fortunately, snowy tree crickets abound in North America, so they're easy to find. The problem is, how do you tell snowy tree crickets from any of the other 1,000 different cricket species? Their name provides a clue, but not how you might think. First, don't look for snowy tree crickets anywhere near snow. Second, they don't hang out in trees. Usually found in leafy bushes, these insect thermometers get their name from their coloration: snowy white as developing nymphs, turning to a pale, treelike shade of green as adults.

If you can see one, you may recognize it not only by this pale-green color but also by its body shape. The snowy tree cricket has a classic teardrop shape, nearly pointed at the front, then tapering to a rounded rear body.

There is one other way to tell, if you can get close enough. Snowy tree crickets don't chirp by rubbing their legs together. Instead, their distinctive "treet-treet-treet" sound emanates when they rub one of their front wings against a raspy vein on the other. So it's easy—just ask one to "treet" for you while you watch.

The Human Factor: Humiture and Windchill

Sometimes it seems that the temperature defines a day and even a whole climatic region. The main element to every weather report, for example, is the temperature. Cloud cover and precipitation are close seconds, but they're not number one. After all, if you hear that tomorrow will be rainy, there's a vast difference in how you'll plan for the day if the temperature is forecast to be 85° F or if it's going to be 32° F.

Though we use the temperature to estimate the day's comfort level and to determine what we wear, raw temperature readings alone often serve only as a rough gauge of how the weather is going to feel to you. Other weather factors dramatically affect how a given temperature will feel. On cold days the wind is the main comfort modifier. On hot days the humidity adds the most influence to how our comfort is affected by temperature.

Weather reporters have developed two additions to their reports that take these modifying factors into account and help us better estimate how to dress for the day's weather. In the winter they use a factor called *windchill*, and in the summer weather reports often include a factor called *humiture*.

Determining Windchill and Humiture

WINDCHILL TABLE

Wind Speed (mph)	Still-Air Temperature (°F)							
	-30	-20	-10	0	10	20	30	40
5	-36	-26	-15	-5	6	16	27	37
10	-58	-46	-34	-21	-9	3	16	29
15	-72	-59	-45	-32	-18	-5	9	22
20	-82	-68	-53	-39	-25	-11	4	18
25	-88	-73	-58	-43	-29	-14	0	14
30	-94	-79	-64	-48	-33	-18	-3	12
35	-98	-82	-67	-51	-36	-20	-5	11
40	-101	-85	-69	-53	-38	-22	-6	10

HUMITURE TABLE

Relative Humidity (percentage)	Air Temperature (°F)							
	75	80	85	90	95	100	105	110
40	74	79	86	93	101	110	122	135
50	75	81	88	96	107	120	135	150
60	76	82	90	100	114	132	149	163
70	77	85	93	106	124	144	161	**
80	78	86	97	113	136	157	166	**
90	79	88	102	122	150	170	**	**

Here's a formula that will enable you to calculate the windchill factor yourself if you get only raw data from a weather-observation station and do not have a windchill conversion chart.

$$WC = 91.4 - (91.4 - T)\left[0.478 + 0.301\sqrt{(W - 0.02W)}\right]$$

Where:
  WC = windchill
   T = temperature in degrees Fahrenheit
   W = wind speed in mph

Note: This formula is not valid for temperatures greater than 92° or for wind speed less than 4 mph.

Windchill-factor readings and humiture conversions affect only how the weather *feels* to us. They do not affect the actual temperatures of objects outdoors. That is, thermometer readings are unaffected by either one. Yet windchill and humiture are important because of how they affect the temperature-regulating functions of our circulatory systems.

So what does this mean to you personally? When the wind is still, your body will heat a very thin layer of air near your skin, which lessens further heat loss. Wind carries this heated layer away and draws more heat from your body, making you feel colder. As wind speed increases, more heat is drawn out, thus lowering your body temperature and causing you to feel even colder. (Body temperatures below 85° F usually are fatal.)

During hot weather your body cools itself by transferring excess heat to the surrounding air. Even on a very hot day your body can cool itself adequately if the humidity is low, because evaporating perspiration lowers your skin temperature. When the humidity is high, though, the evaporation process slows. This hampers your body's cooling system and you retain more heat, thus elevating your body temperature and making you feel hotter. (Body temperatures above 108° F usually are fatal.)

The sidebar on p. 87 contains tables that show you how to calculate both windchill and humiture. The corrected temperatures that you read from these tables will help you decide what to wear each day. For example, select winter clothes that minimize the heat that a strong wind can carry away. Select summer clothes that will help your body cool itself and maintain its temperature.

The temperature, of course, affects far more than your skin. Temperature is so highly significant that an unusual temperature pattern in one part of the world can have an impact on global weather patterns. Let's look at one of the most significant and famous temperature events anywhere.

Ladybugs

When ladybugs swarm
Expect a day that's warm.

Ladybug beetles are covered by a set of hard wings that hold in body heat. They fly about as a means of cooling off when the weather begins to get warm. Because they do not have a constant internal body temperature, all insects respond to warm weather and fly more readily as temperatures rise. Found especially in the southern United States, and in the Appalachian Mountain regions, this little poem is actually a fair predictor of weather.

Global Temperature Effects on Local Weather: El Niño

El Niño—which means "the child" in Spanish because it usually reaches its peak around Christmas—is basically a change in the flow of warm- and cold-water currents in the Pacific Ocean. This major climatic change, occurring every three to five years, can generate all kinds of devastating weather, from droughts in Africa to flooding in North America. El Niños sometimes cause heavy rain in the southeastern United States and can make the hurricane season there milder than usual. They warm up western Canada.

"Apart from the change in seasons, El Niño is the most important recurring event affecting world climates," says David Rodenhuis, director of the National Weather Service's Climate Analysis Center in Camp Springs, Maryland.

Normally, cool water moves from Antarctica north along the west coast of South America. That current then turns due west along the equator and continues on into the central Pacific Ocean. When an El Niño cycle is in progress, this cool water current weakens, allowing warmer water over the central Pacific to move east and replace the cool water normally found along the western South American coast. In the 1982–1983 El Niño, the sea surface temperature rose as much as 14° F above normal while in the 1991–1992 occurrence, the average surface readings climbed only about 4° F above normal.

The change in water temperatures off South America has a rippling effect on the weather and changes jet-stream patterns worldwide. Meteorologists have just begun to understand what weather patterns are altered during an El Niño. Experts say that the 1991–1992 El Niño set the stage for the heavy rains that inundated the U.S. Gulf Coast and California during that season, causing disastrous flooding. In the north it delivered above-normal warmth to western Canada and to northern states from Washington to Minnesota. Indonesia and Australia were drier than usual during the same period.

The most notable and beneficial effect of an El Niño comes with hurricane development. Developing tropical storms are usually ripped apart by strong jet-stream winds in the upper levels of the atmosphere. This normally prevents most disturbances from building into full-blown hurricanes. In fact, the strong El Niño years of 1983 and 1987 were very quiet hurricane years.

Weather patterns in the United States also are affected, sometimes drastically. During an El Niño year, the East and West coasts experience unusually stormy weather. The Ohio Valley usually gets away with a mild winter with near-normal precipitation. Occasional snows and cold waves, however, can occur. The following summer usually features dry weather in the southeastern United States and the Ohio Valley. By the following winter, as the El Niño subsides and the currents return to their normal positions, the two coasts have fairly quiet winters, while the central United States receives a cold and snowy winter. This was certainly the case in 1984, when many cities in the Midwest received record snows and frequent blasts of Arctic cold.

El Niños occur about once every four years and can vary greatly in intensity. The 1983 El Niño was the most pronounced in decades, while the 1987 El Niño was relatively mild. The 1993 El Niño was a moderate one.

El Niños and their effect on world climate have been studied only during the past fifteen years or so, but they have a long history. (See the article below.) Much more needs to be learned about them. This cyclical change in a simple ocean current has taught us a very important lesson: Earth is a fragile system of checks and balances, and any small change in one of those systems can cause radical changes in others. Such is the case with "the child."

El Niño Watch Advisory

A monthly advisory concerning El Niño conditions on the western coast of the United States is available free via U.S. mail or fax from NOAA's Coastal Ocean Program, CoastWatch, El Niño Watch, at the following address:

> U.S. Department Of Commerce
> National Oceanic and Atmospheric
>   Administration
> National Marine Fisheries Service
> Southwest Fisheries Science Center
> P.O. Box 271
> La Jolla, CA 92038–0271
> (619) 546-7613
> fax: (619) 546-5614

The text provides information concerning current conditions as well as information having to do with conditions in previous months. It will also give you a capsule summary of how the El Niño is affecting marine life, both in general and by specific species. When you make your request, ask for all the back issues. In a few days you will receive a package that will bring you up to speed on the current event.

The World's Oldest "Child"

[In 1991] more than 300 Weather Service buoys, loaded with electronic equipment and floating around the equatorial Pacific, gave scientists their earliest-ever advance warning of an El Niño.

Information collected from small, drifting buoys that monitor temperatures near the ocean's surface and from large, moored buoys that measure currents and subsurface sea temperatures is beamed to satellites.

"When El Niño hit in 1982, we only had a dozen or so of these things out there," says Richard W. Reynolds, an oceanographer at the [National Weather Service Climate Analysis Center]. "They've really made a difference in the accuracy and confidence of our predictions."

Encouraged by the early forecast in 1991, meteorologists hope that they soon can develop models of oceanic and atmospheric conditions that start the weird weather pattern.

"Linking these to a computer," says Vernon E. Kousky, a research meteorologist at the center, "we someday hope to predict an El Niño even before the first symptoms appear...."

El Niños have been around for at least 2,500 years, concludes Miriam R. Steinitz-Kannan, a biologist at Northern Kentucky University, whose work has been supported by the National Geographic Society. She made her findings after examining and dating deep sediment cores from an Andean lake in a region of Ecuador that is often hard hit by El Niños.

The cores taken from the Andes site and from lakes in the Galapagos Islands off the coast of Ecuador show that freshwater diatoms, a type of microscopic algae, thrive in such places during and immediately after El Niño years because heavy rainfall displaces the normally salty water.

"So far it looks as if severe El Niños have occurred less frequently in the last 500 years," Steinitz-Kannan says.

Trigger mechanisms for El Niños usually come into play during the spring or autumn, when world weather systems are shifting.

"There are two of these features that dominate the tropics and subtropics during crucial times of the year," Kousky explains. "One is the monsoon that dominates Southeast Asia and India during our summer, and the other is the Australian monsoon that influences that region during our winter.

"Niños seem to evolve when these two monsoons are in transition, moving cloudiness and precipitation around during our spring and fall. And true to form, we first noticed the symptoms of the current one last spring and were able to confirm it by November."

—By Donald J. Frederick.
Reprinted with permission from the National Geographic Society News Service.

One of this book's authors, Ron Wagner, had the "good" fortune to be in Pennsylvania when the all-time record cold temperature was recorded. Extremely low temperatures not only have a dramatic effect on humans, but on machines as well. Here is Ron's account of that day:

I was flying Boeing 727s for Eastern Air Lines and spent the night in Pittsburgh when the city experienced its all-time record low temperature. We had an early morning departure, and I had the unfortunate "honor" of preflighting the outside of the aircraft with a wind chill factor of -85° F.

I started the aircraft's auxiliary power unit and started the heating system. The ground crews helped out by attaching supplemental heating hoses. After a couple of quick cockpit checks, it was time to do the outside walk-around.

Fortunately, I was dressed properly, with thermal underwear, thermal ski gloves, and a hat that covered my ears. The flight attendants lent me a couple of scarves and stuffed them in any gaps they could find. Another flight attendant lent me some baggy sweat pants that I pulled on over my uniform pants. The best deal for me was that the captain was a huge man and his overcoat fit over me after I had bundled up with my own overcoat.

All those layers of clothing really did the trick. That was important because at such a cold temperature, I couldn't hurry the preflight—it was more important than ever to check the aircraft thoroughly for leaks or signs of ice. But everything went well and I survived the outside preflight without much suffering.

I returned to the cockpit and found that the plane was beginning to warm nicely and the captain was in his seat, running his checklist. After stripping off layer upon layer of clothing I took my seat. We were all surprised that everything on the aircraft checked out normal, but we were very excited that in a few minutes we'd be on our way to Florida for a Tampa layover that night. Boy, we were ready for that!

Then it happened. A sight I'll always remember.

I was talking to the captain when I saw cracks quickly spread all across the side window right behind him. In a few seconds, the window was a crazy quilt of cracked lines and we couldn't see through it. The stress of heating quickly to a comfortable temperature from such an extremely cold night was too much.

The flight was canceled and we waited six hours for a new window to be flown in and the replacement to be made. By then we just had to ferry the plane to Tampa and that afternoon I was floating on my back in a swimming pool under sunny skies with a temperature of +86° F—a change of 171° F in just a few hours!

Highest and Lowest Recorded Temperatures

In chapter 5, you learned why the equator is not the hottest place on the planet, though you might expect it to be the hottest, since the sun is much stronger in equatorial regions. The wind carries the hottest air away from the equator, so it has its greatest impact on other latitudes. Here's a summary of U.S. and world temperature records so you can see how temperature is distributed around the planet.

High-Temperature Extremes

- **Hottest average summer temperature:** Death Valley, California. Daytime highs in the summer can average 116° F, with overnight lows often remaining above 100° F.
- **Hottest average year-round summer temperature:** Dallol, Ethiopia. Year-round mean high temperature is 94° F.
- **Hottest average year-round U.S. summer temperature:** Key West, Florida. Year-round average high temperature is 78° F.
- **Longest recorded hot streak:** Marble Bar, Australia. Between October 30, 1923, and April 7, 1924—a string of 162 consecutive days—the temperature never dropped below 100° F.
- **Hottest recorded hot streak:** Death Valley, California. Between July 6 and August 17, 1917—a string of forty-three consecutive days—the daily high temperature reached 120° F or higher.
- **Hottest recorded temperature:** El Azizia, Tripolitania, Libya. On September 13, 1922, the peak temperature hit 136° F.
- **Hottest recorded U.S. temperature:** Death Valley, California. On July 10, 1913, the peak temperature hit 134° F.

Low-Temperature Extremes

- **Coldest average U.S. winter temperature:** Point Barrow, Alaska. Winter temperatures here average –9° F.
- **Longest U.S. cold streak:** Havre, Montana. Once experienced below-zero temperatures for nearly seventeen days in a row.
- **Coldest recorded temperature:** Vostok, Antarctica. On July 21, 1983, the temperature dropped to –129° F. (Accurate temperatures were not recorded for Antarctica before 1957. At Pleasteau Station in Antarctica, the year-round mean temperature is –70° F.)
- **Coldest recorded temperature on a populated continent:** Oymyakon, Siberia, Russia. The coldest temperature ever recorded was –90° F on February 3, 1933.
- **Coldest recorded U.S. temperature:** Prospect Creek, Alaska. On January 23, 1971, the temperature dropped to –80° F. (Within the forty-eight contiguous states, the all-time low was recorded as –70° F at Rogers Pass, Montana on January 20, 1954.)
- **Most widespread U.S. cold wave:** February 1899 was so cold that the Mississippi River froze over its entire length, with ice as thick as 2 inches even in New Orleans.

U.S. Temperature Records

State	High (°F)	Low (°F)
Alabama	112	–27
Alaska	100	–80
Arizona	128	–40
Arkansas	120	–29
California	134	–45
Colorado	118	–61
Connecticut	106	–32
Delaware	110	–17
Florida	109	–2
Georgia	113	–17
Hawaii	100	7
Idaho	118	–60
Illinois	117	–35
Indiana	116	–36
Iowa	118	–47
Kansas	121	–40
Kentucky	114	–37
Louisiana	114	–16
Maine	105	–48
Maryland	109	–40
Massachusetts	107	–35
Michigan	112	–51
Minnesota	114	–60
Mississippi	115	–19
Missouri	118	–40
Montana	117	–70
Nebraska	118	–47
Nevada	125	–50
New Hampshire	106	–46
New Jersey	110	–34
New Mexico	122	–50
New York	108	–52
North Carolina	110	–34
North Dakota	121	–60
Ohio	113	–39
Oklahoma	120	–27
Oregon	119	–54
Pennsylvania	111	–42
Rhode Island	104	–25
South Carolina	111	–19
South Dakota	120	–58
Tennessee	113	–32
Texas	120	–23
Utah	117	–69
Vermont	105	–50
Virginia	110	–30
Washington	118	–48
West Virginia	112	–37
Wisconsin	114	–54
Wyoming	114	–63

Autumn Foliage Colors

Indian myth tells us that heavenly hunters killed the Great Bear of the heavens in the fall and its blood dripped over the forests, coloring some of the leaves red. Other leaves turned yellow when fat splattered out of the celestial caldron as the hunters cooked the meat.

The United States Department of Agriculture takes a less romantic view: In spring and summer, leaves manufacture food for the trees in cells containing the green pigment chlorophyll.

CHAPTER 6: TEMPERATURE

Leaves also contain the same yellow-orange substance that gives carrots their color, but the greater amount of chlorophyll makes them appear green.

In the fall, lower temperatures begin to stop the food-making process and the chlorophyll breaks down so that the yellow-orange colors emerge.

At the same time, cool nights trap sugar in the leaves, forming a red pigment. Differing amounts of red and yellow pigment account for the wide variety of colors in autumn leaves.

Exposure to strong, fall sunlight tends to bring out the brilliant colors, while prolonged cloudy, fall weather brings out the pastel shades. Here is a list of some of the different types of trees and the color of their leaves in autumn:

Scarlet oak—red	Black oak—yellow
White oak—reddish to violet	Beech—yellow
Silver maple—pale yellow	Mountain maple—bright red
Striped maple—yellow	Red maple—red
Aspen—yellow	Birch—yellow
Hickory—yellow	Black gum—scarlet
Mountain holly—yellow	Smooth sumac—red
Dwarf sumac—purple to red	Black chokeberry—bright red

Unlike most of the trees whose leaves turn colors in autumn and then fall off, the leaves of the oak tree eventually lose their bright colors, turn brown, and remain on the tree most of the winter. They fall off very slowly during the course of the winter, in strong winds. For the latest information on next fall's foliage viewing, call these state hotlines for details:

National Forest Service	1-800-354-4595
Northeast:	
Connecticut	1-800-282-6893
Maine	1-800-533-9595
Massachusetts	1-800-632-8038
New Hampshire	1-800-262-6660
New York	1-800-225-5697
Rhode Island	1-800-556-2484
Vermont	1-800-VERMONT
Mid-Atlantic:	
Pennsylvania	1-800-FALL-INPA
New Jersey	1-800-354-4595
Delaware	1-800-441-8846
Maryland	1-800-532-8371
Virginia	1-800-434-LEAF
North Carolina	1-800-847-4862
Midwest and West:	
Kentucky	1-800-225-8747
Michigan	1-800-644-3255
Minnesota	1-800-657-3700
Oregon	1-800-547-5445
Tennessee	1-800-697-4200
Washington	1-800-354-4595
Wisconsin	1-800-432-TRIP

—Courtesy of Stormfax Weather Services, (610) 66STORM

Temperature Products

Galileo Thermometer

The Nature Company
Catalog Division
P.O. Box 188
Florence, KY 41022
(800) 227–1114

Elegantly sculpted in glass, this unique thermometer recreates an actual seventeenth-century experiment by the astronomer Galileo. This instrument uses handblown glass spheres of different weights that float upward or downward, depending on temperature, in a 24-inch-tall glass tube filled with water. Each glass sphere carries a number that corresponds to the temperature of the water when the respective sphere ball floats. Called the Termometro Lento, this is a beautiful, contemporary interpretation of a classic science experiment. At $259, it's not cheap, but it is an exquisite conversation piece.

Nimbus Temperature Monitor

Sensor Instruments Company, Inc.
41 Terrill Park Drive
Concord, NH 03301
(800) 633–1033
In New Hampshire: (603) 224–0167

As with all Nimbus instruments, this model is packed with features: thirty-five-day and hourly memory, measurements from –40° F to +140° F at an accuracy of ±1° F, switchable to degrees Centigrade and Kelvin on a large LCD display. Its high/low memory will help you track weather patterns and trends and maintain a long-term temperature log. You can have your choice of a cherry or oak case. The unit is powered by four "C" cell batteries (included) and carries a three-year warranty. It

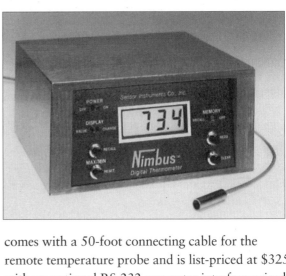

comes with a 50-foot connecting cable for the remote temperature probe and is list-priced at $325, with an optional RS-232 computer interface priced at $100.

Two Compact Digital Models

Robert E. White Instruments, Inc.
34 Commercial Wharf
Boston, MA 02110
(800) 992–3045

White offers two versatile digital thermometers with unique features. One is the Taylor #5566 "Digital Humidiguide," for carrying on trips or for checking conditions from room to room. It simultaneously presents ambient temperature and humidity on a large LCD display. Temperature is in degrees Fahrenheit or Centigrade in 0.1° increments. Humidity is relative, in 1-percent increments. Powered by two "AAA" batteries, it's only 5½ inches long and ¾ inch thick for true portability, but it can be wall mounted. The Taylor #5566 is priced at $63.

The Taylor #5395 comes with a rugged temperature probe on a 36-inch wire. It is designed to measure temperatures in liquids, air ducts, and so on. In addition to the probe temperatures, it presents both ambient air-temperature and time in a large LCD display. It comes in a shirt-pocket size (4 by 2¼ by ¾ inches), is powered by one "AA" battery, has a temperature range of 0° F to 160° F, and costs $31.98.

CHAPTER 6: TEMPERATURE

Computemp
Plow & Hearth
301 Madison Road
Orange, VA 22960–0492
(800) 627–1712

Replace your standard bedside alarm with this digital wonder that adds the functions of a recording thermometer to an LED clock. Use the included 30-foot cord to connect the unit's outdoor sensor and it will display both indoor and outdoor temperatures, recording the highs and lows of both along with the times they occurred.

Perhaps Computemp's most valuable feature is its programmable temperature alarm, which will alert you when your limit has been hit—perfect for the gardener who's worried about a frost. It's housed in an attractive black ABS plastic case with wood-grain finish and costs $79.95.

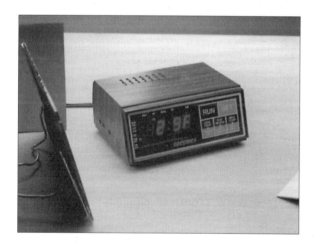

In/Out Digital Thermometer
Plow & Hearth
301 Madison Road
Orange, VA 22960–0492
(800) 627–1712

Check your indoor and outdoor temperatures with digital accuracy using this attractive, compact thermometer. It includes an external sensor on a 10-foot cord with double-stick tape for quick mounting. Its high-tech design features large, easy-to-read numbers and doubles as a digital clock. It is priced at $24.95.

Casio Recording Thermometer/World Time Watch
Wernikoff's Jewelers
2731 N. Milwaukee Avenue
Chicago, IL 60647–1388
(800) 932–8463

Casio makes an electronic, digital watch—Model #WWT2—that includes a temperature sensor and computer to measure current temperature and track temperature trends. As a watch, the WWT2 displays time zones around the world and offers five time alarms, a twenty-four-hour countdown timer, a calendar, a twenty-four-hour stopwatch, and an hourly time signal. As a temperature computer, it measures in both degrees Fahrenheit (from –14° F to 140° F) and Centigrade (from –10° C to 60° C). World travelers will appreciate its temperature history: It lists average high and low temperatures around the world by month. It will store local temperatures in memory, on the hour, for twenty-four hours and has a temperature alarm. It comes with a one-year warranty and a two-year battery and is priced at $59.98, plus $5.00 shipping and handling.

Temperature Recorders
Robert E. White Instruments, Inc.
34 Commercial Wharf
Boston, MA 02110
(800) 992–3045

You can get just about anything you want in a temperature recorder, but here are four models that are good choices for a wide variety of users: Two

record temperature only and will serve most users. Two record both temperature and humidity and meet professional demands.

The Taylor #2354 Electric Thermograph is a great value. Mounted on a black plastic base, it will chart temperatures around the clock for seven days. Powered by 110-volt AC, the unit can record temperatures from –40° F to 120° F and has a remote sensor with a 40-foot cable. It includes charts, pen and ink, clear plastic cover, and instructions. It is priced at $440.

The Maxant #65 Compact Thermograph is nearly a clone of the Taylor model, including price, but it operates on an "AA" battery for portability.

Museums exhibiting a Renoir, or growers cultivating prize-winning flowers, demand the best equipment. The Belfort #5-594 Hygrothermograph is the industry standard and attains its superb accuracy using a gold-plated Bourdon tube temperature sensor and a hair-humidity sensor. It's available in two portable models, one battery-powered, the other powered by a windup key. Both record temperature and humidity for seven days and come with charts, pens, ink, and an instruction book. Call White for pricing of this premier professional model.

Another professional-quality hygrothermograph is the Maxant #173. Very similar to the top-of-the-line Belfort—minus the gold-plated probe—this model runs on an "AA" battery for portability and includes all standard accessories. An added feature is its locking case, which maintains security. It's priced at $600.

Chapter 7
Moisture

Without moisture in the atmosphere, nothing that we call weather would exist. Moisture is the principal component of all weather events, from morning dew to hurricane rains. Even when there seems to be no active weather, the moisture in the air still affects us through humidity levels, which can bring cooling comfort or cause sweltering misery. Moisture's state is the prime factor in how it is classified in weather.

Atmospheric moisture exists in three states:

- *solid*—ice, hail, frost, snow, or sleet
- *liquid*—dew or rain
- *vapor*—humidity, clouds, or fog

The sun's energy heats the Earth's atmosphere unevenly. This results in a continual imbalance in the heat distribution within the atmosphere that the laws of nature work full time to even out through heat exchange. This constant equilibrium-seeking heat-exchange process initiates three different mechanisms that affect the state of atmospheric moisture. Most of the action of weather involves one of the following mechanisms as they transform moisture between its three states:

- *Evaporation:* Changes liquid moisture into its vapor (gaseous) state. The rate of evaporation increases as temperature increases, because hotter molecules move faster. Faster-moving molecules escape from the surface of the water more freely to become water vapor. Air pressure also affects the rate of evaporation.
- *Condensation:* This is the reverse of evaporation, changing water vapor into its liquid form. Air has a limit to the amount of moisture it can hold. The limit is affected by temperature, and air will hold less moisture as it cools. Once air temperature drops below its saturation point, the excess moisture turns back into liquid form.
- *Sublimation:* As ice (a solid state of moisture) is heated, it normally turns to liquid, then to gas. Under certain conditions, however, moisture can pass directly from a solid into a vapor, or vice versa. Snow is a result of a sublimation process that turns water vapor directly into solid form, without first becoming a liquid.

Measuring Moisture Content

One measure of the amount of moisture in a particular air mass is its *dew point*—the temperature to which air must cool to reach its *saturation point*. Saturation is the point at which air, at a given temperature and pressure, is holding its maximum possible amount of moisture. If the air over a body of water is at its saturation point, then evaporation stops until a pressure or temperature change causes the saturation point to change. If the saturation point increases, evaporation will resume. If it decreases, condensation will take place.

Here's an example. If a certain air mass is at 72° F and its dew point is at 65° F, then the air will reach saturation if it cools to 65° F. If cooling continues the air no longer will be able to hold the moisture it contains, and some will come out in the form of condensation.

Humidity also can be measured using an absolute scale, called *specific humidity*, to express how much moisture a given air mass can contain at its current temperature and pressure. This figure would be of little use without some technical knowledge of the air's capacity to hold moisture, however, so weather observers use specific humidity readings as a factor in calculating the more familiar *relative humidity*. Air with a relative humidity of 100 percent is at its saturation point. Thus we have an easy-to-understand number that expresses how close the air is to being saturated.

Precipitation Types

When the temperature or the pressure of an air mass causes it to reach saturation, the moisture that it no longer can carry usually comes out in one of the familiar forms of precipitation. Here's a brief listing of each type:

- *Rain:* Precipitation that reaches the earth as large droplets. It may begin as smaller droplets high in the sky, but as the droplets fall, they intermingle and combine to become larger droplets. We classify rain by the amount that falls and by the results of the free-fall mixing action, describing rain as *light, moderate,* or *heavy.*
- *Drizzle:* The precipitation droplets from stratiform clouds are much smaller than those from other forms. Stratiform clouds have little turbulence action. Thus, the droplets do not combine and are still small when they reach the ground.

Dorchester, Massachusetts, after a snowstorm

- *Freezing rain* and *freezing drizzle:* During certain conditions precipitation becomes supercooled and a thin coating of ice forms over the droplets. When they reach the ground, they coat the ground with ice.
- *Snow, sleet, ice pellets,* and *hail:* In these forms of precipitation, moisture has frozen solid. Ice pellets and sleet begin falling as normal rain but freeze upon passing through a layer of below-freezing air. Hail occurs when updrafts carry these frozen pellets back up through the freezing layer to pick up more moisture, then come back down for the freezing air to add another layer of ice. Eventually, the pellets become large enough that the updraft cannot lift them again, and they strike the ground still frozen. Snow forms when sublimation causes moisture to pass directly from liquid to solid.

There's more to moisture, though, than these obvious, visible precipitation forms. In fact, moisture is all around you, no matter where you are.

Water, Water Everywhere

A fiercely competitive Little League baseball game comes screeching to a halt when the clouds above open up and turn the playing field into a mud bath. That same rain shower may bring life back to a farmer's parched crops. All living things depend on and need rain—although that's hard to explain to a team of disappointed ten-year-olds.

Even under the blazing sun over an arid desert, the air is rich with water. You can see it all the time. Even on a bright sunny day, you may still see a haze that makes distant objects look dull and hazy. You may believe that you are seeing a sky obscured by pollution, but this haze has been around as long as the hills it obscures. Dust also fills the air and is a natural ingredient that is necessary for weather to occur.

Moisture—drawn into the air from rivers, lakes, and oceans and transpired by trees and other plants—literally fills the atmosphere. Though it is ever-present, most moisture is not concentrated enough to become visible. Only when the air cools enough will its moisture condense onto the airborne particles of dust, salt, and smoke to become visible droplets that form clouds and haze.

There are "warm" clouds—the temperature is above freezing—and "cold" clouds, which are formed by billions of droplets and tiny ice crystals. Drizzle falls from the warm clouds when many droplets come together and become too heavy to float.

The cold clouds hold ice crystals as well as water droplets; so by the time they reach the ground as either rain or snowflakes, they have grown quickly by collecting any droplets or small crystals that were in their path as they fell.

Rain falling from the sky is partially acidic, because it dissolves gases from the air. Near industrial areas the acid rain is more prominent, although rain is never pure. Muddy residue is left after a rain when the drops have collected dust suspended in the air. While we all must work to keep our air as clean as possible, totally clean air could not produce clouds nor rain.

Rain gauges collect the water that falls and allow weather observers to measure the amount. Light rain, or drizzle, falls from layered clouds less than 1.2 miles thick as small droplets of less than 0.02 inch across and take at least an hour to hit the ground. Moderate rain falls at .02 to 0.16 inch per hour. Heavy rain falls from a stacked cumulonimbus cloud, which may be 9 miles or more deep. A heavy rain in the Indian Ocean once brought the island called Cilaos 74 inches of rain in twenty-four hours. At Smithport, Pennsylvania, on July 18, 1942, 30.8 inches of rain fell in four and one-half hours; and in 1969, Mount Washington, New Hampshire, received more than 130 inches of rain.

There are three main types of rainfall: convectional rain, relief rain, and frontal rain. *Convectional rain* falls all year round near the equator and during the warmer months in other parts of the Earth. The sun's heat warms the land and the sea, causing the moist air to rise into the sky. As it cools it spreads, forming the cumulus—or convection—clouds that bring the convectional rainfalls.

Mountains receive *relief rain* as the air blowing across the vast oceans picks up moisture. This moist air must rise to cross the mountains. The rising air cools, and rain falls on the mountains' coastal slopes. By the time the air descends on the far side of a mountain, it is much drier, and so is the climate on that side of the mountain.

A *frontal rain* is the most important source of rain in many places of the world, affecting hundreds of square miles in a single day. Heat is blown from the tropics to both poles by wind, and in the middle latitudes, the returning cold winds meet up to form a polar front. Depressions, or low-pressure systems, often happen along this front because the cold air forces the tropical air to rise, pulling it into a depression. A heavy, steady rain follows before the depression fills and the clouds disperse.

Subtle Precipitation Forms: Dew, Fog, and Mist

On a chilly morning you quickly prance barefoot across the lawn to retrieve the newspaper and duck back into the house. But your feet get soaking wet, even though it hasn't rained. Why is that?

Not all of the air's moisture forms into droplets on airborne particles. Some forms over objects on the ground and on the inside surface of windows. The tiny droplets of water sparkling on the lawn and the cars in the morning are *dew*, which forms when moist air reaches objects that cool the air to below the dew point (the temperature at which condensation occurs).

During the day the sun's heat is absorbed by such surfaces as grass and cars. These surfaces then warm the nearby air, causing moisture on them to evaporate. After nightfall during clear, calm weather conditions, the grass and cars lose heat, chilling the nearby air to below its dew point. Dew forms even more readily on grass and plants because they con-

Fog Rising from a River

tain moisture that is added to the moisture in the air.

Dew usually forms only on still nights without clouds. On a windy night the cooling effect is dispersed, so that air near the ground does not reach its dew point. On overcast nights, clouds act like a vast blanket holding heat near the ground so the air remains above its dew point.

Another type of dew—called *advection dew*—forms when moist, warm air comes in contact with cold objects that cool nearby air to a temperature below its saturation point. This is what you see on your bathroom mirror after taking a hot shower.

A fog cloud is very similar to the steam that billows out of a rapidly boiling tea kettle—the hot moisture saturates the air and condenses in the cooler air outside. A million tiny droplets of water form a cloud on either land or water, causing a fog. Fog over land is most common in the winter and fall, when the nights are longest and the air closest to the ground cools below its dew point.

Valleys are the first areas to be overtaken by fog because the coldest, heaviest air drains here, where the air is already filled with moisture evaporated from lakes, rivers, and vegetation. *Radiation fog* occurs when slightly blowing wind spreads the cooling effect and the dew point is reached by more air, causing the fog to become more widespread.

Advection fog occurs, just like advection dew, when warm, moist air crosses a cold surface. Warm air blowing across a cold ocean current causes a sea fog. Fog forms inland over a frozen lake in much the same way.

In cold weather it sometimes looks like smoke is rising from water. This eerie fog is *arctic sea smoke*, which occurs when cold air moves across warmer water, mixing the cold air with the thin layer of warm, moist air near the surface.

Most fogs are not deep, so at sea a ship's masts often appear above the low-lying fog. A very thin fog—one having very few water droplets—is called a *mist*.

Fog droplets can freeze on any object they touch when the temperatures dip to below the freezing mark. This is called *freezing fog*. The frost often remains a long time after the fog has cleared.

Now let's look at another form of frozen precipitation, one that is neither so subtle nor as picturesque as a fragile coating of morning frost.

What the Hail?

Hail is a special case of precipitation. It's a mixed-up combination that results when a couple of factors are just right. It begins as regular rain

Hailstones

that freezes as it falls through a layer of subfreezing air. If these frozen drops continued to fall to Earth, they would be called *sleet;* but instead, strong updraft winds fling these frozen raindrops back up in the air high enough to pick up more water, then fall back through the freezing layer, which freezes the new coating. The updraft then carries the drops back up, repeating the process over and over. Eventually, the growing hailstone becomes too heavy to be carried back up and completes its fall to Earth.

The stronger the updraft winds, the more return trips hailstones will make and the larger they will become. You can tell how many return trips a hailstone made by cutting it in half and counting the distinctive layers of ice.

Hailstones can grow to be quite large, often reaching golf-ball size. Large hail damages cars, buildings, and crops and can cause injury to persons caught outside. The largest hailstone recorded fell in Coffeyville, Kansas, on September 3, 1979. It measured 17.5 inches in diameter!

All about Raindrops

You easily can determine the size of raindrops through a simple experiment. All you need is a dry piece of cardboard and a finely calibrated ruler. Place the cardboard outside and let a few raindrops hit it. You'll find they measure from 0.01 inch to more than 0.25 inch across. Remember that each measurable raindrop actually is the result of millions of droplets of water combining into a single unit.

Rain less than 0.02 inch across is classified as *drizzle*. These tiny drops are so small and light that they may take more than an hour to fall to earth. Drizzle usually falls from thin, layered clouds less than 1.5 miles thick. Such thin clouds give the droplets little chance to intermingle and combine to form larger rain drops.

A heavy shower, such as occurs in a thunderstorm, can produce large raindrops. Perhaps more than 10 miles thick, tall cumulonimbus clouds give the droplets plenty of time to combine into raindrops of up to 0.25 inch across.

There is a big misconception about the shape of raindrops. If asked to draw one, most people would sketch a classic teardrop, streamlined shape. However, that falling raindrops could take on this shape doesn't make any sense when you consider the friction of air through which they fall. Air presses against the bottom of speeding raindrops, causing them to flatten out into the shape of a hamburger—relatively flat across the bottom and rounded on top.

Water: The World's Oldest Civil Engineer

Running water works at eroding away the land, leaving it changed forever by cutting valleys through mountains and carrying the broken rock down to the oceans. As the rivers meet up with the seas, they shed their load of rock. That same rock starts to build new land.

Steep mountainsides are the first to be eroded by running water. At the peaks of mountains and hills, rivers and springs race downhill, picking up loose pieces of rock, which hammer against the streambed, cutting a valley into the mountain. This constant pounding chips away at the stones, turning them into gravel and then sand and mud, also deepening the bedrock of the stream.

When the softer rocks downstream are worn away faster than the very hard rock, a waterfall is created. A stream can become a raging river when tributaries flow into it from either side, leaving numerous valleys and ridges. After thousands of years, a mountain has been reduced to low ridges and gently sloping valleys.

A lowland river wears away at the land along the bank instead of into the riverbed. It eats away at the sides, where the current is strongest. The river rushes from side to side in huge loops that are called *meanders* because the mud removed from one bank often accumulates on the other side. Meanders widen and flatten the valley floor as they work their way downstream.

Some wide, slow-moving rivers carry the smallest amount of clay and silt and flow across a huge, low river plain. Eventually, they just dump their load into the sea.

The silt may pile up, building new land along the shallow, protected coast, as it does on the Mississippi River. The Mississippi finally reaches the sea after passing through the low, triangular-shaped silt plains, called *deltas*, that they have built far from the mainland.

Water also attacks and permanently changes the land in the form of ice. The last ice age lasted for hundreds of thousands of years and covered much of the North American continent with ice. As it melted the moving glaciers simply erased entire mountain ranges, flattening them into low plains. Slow-moving glaciers—inching forward almost imperceptibly—are still carving the Earth today in some mountain ranges.

Glaciers also attack and change the land by grinding away at rocks on many high mountains. When snow falls and is packed into cavities high upon the mountains, a glacier is beginning to form. Eventually, the snow at the bottom turns to ice from the weight of the snow that continues to pile on top. This constant pressure finally causes a piece of ice to jut out and begin to roll, very slowly, downhill.

The glacier follows the path already chiseled away by the river, picking up pieces of broken rock along the way and further breaking away rocks, making the valley very deep and wide.

At the same time the rocks around the hole where the glacier began crumble away because of the continual freezing and then melting. The hole grows larger and eats away at the mountain. If one or two or more glacier holes are growing at the same time, they eventually will almost join, leaving only a ridge between them.

In cold climates *icebergs* are formed when the glaciers finally make their way downhill to the ocean and chunks snap off and float away. In warmer climates the glaciers make it only halfway down the mountain before they melt and form an *end moraine*—a pile of stones that have dropped off and created a dam that traps runoff water. The narrow "finger lakes" in upstate New York as well as some in the Alpine valleys were formed by this type of natural rock dam.

Clay, gravel, sand, and rock are spread by the melting water from a glacier. Even mighty boulders have been carried long distances by moving sheets of ice. Of course something on that scale takes place only over centuries or, perhaps, millions of years. In such a vast and ancient system of earth-moving precipitation patterns, our observation records are relatively speaking but a blink of an eye. In this blink of an eye, though, we've collected some amazing precipitation records. As you read the following world precipitation records, just imagine what they would be if we had records on the last couple of million years!

World Precipitation Records

Here's a collection of various precipitation records gathered from around the world:

- **Greatest single-day rainfall:** 73.62 inches on March 15, 1952, at Reunion Island in the Indian Ocean
- **Greatest single-day snowfall:** 75.8 inches on April 15, 1921, at Silver Lake, Colorado
- **Greatest single-month rainfall:** 366 inches (more than 30 feet!) during July 1961 at Assam, India

- **Greatest single-year rainfall:** 1,041 inches from August 1880 to August 1881 (watch out for those upside-down years!) at Assam, India
- **Highest average annual rainfall:** 460 inches at Mount Waialeale, Kauai, Hawaii
- **Lowest average annual rainfall:** 0.03 inch at Arica, Chile
- **Longest recorded rain-free period:** more than 400 years in the Atacama Desert, Chile
- **Greatest snowfall from a single storm:** 189 inches from February 13 to February 19, 1959, at Mount Shasta, California

Rain Storm

"Those lazy cloud people don't want to work!" grumbled the Zia Indians whenever thunder and lightning came before the rain. They said the rumble of the Thunder's flapping wings and the flaming arrows shot by the Lightning People were meant to frighten the Cloud People into making rain for the parched earth.

The Seneca were the largest tribe in the Iroquois Confederacy. They lived in the area of northern New York State where Lake Ontario and Lake Erie come together. Among the Seneca, women had great power. Chiefs and tribal council members were nominated by women, and the women removed them if they misbehaved. Here's a Seneca tale about rain storms that illustrates the great power Seneca women held in their society.

Jijogweh, the evil monster bird, made the storms and lived on the blood of humans. The terrified people never knew when the wandering, bloodthirsty, evil bird would swoop down out of the clouds.

The people crept about carefully and watched the skies, but Jijogweh was as cunning as he was swift and powerful. Sweeping his gigantic wings downs along the lakes and rivers, he made the waters churn and hiss until clouds boiled up to hide him. Then with one touch of his wings, the clouds turned angry black, pouring rain and slimy snakes on the poor, frightened people.

There were some people who found the courage to shoot their arrows at Jijogweh when he flew at them, but non of them lived to tell of it. If his poison breath hadn't killed them first, their arrows fell back, broken as Jijogweh attacked. And should even one feather fall from his wing, the blood drops turned to cold rocks that smashed those still alive below.

One night as all the people slept, a dream message came to a strong, young woman named Brave Girl. She was told to shape a strong bow from the wood of an ash tree and string it with her own long, black hair. And she was to feather her arrows with down from the breast of a young eagle. She was told that her arrows shot from this bow would kill Jijogweh.

The young woman left at dawn and climbed a rocky cliff to a high eagle's nest. She dropped bits of meat into the baby birds' open mouths and plucked feathers from their breasts as they looked up to eat.

She hurried home and made her bow and arrows as she had seen them in her dream. Then she went to her tribe's medicine man. He gave her a pouch of sacred tobacco to hang around her neck. Before she left, he asked the good spirits to help her and give her sacred words for guidance.

She waited until dark, feeling safe because of the preparations she had made and the blessings she had received. Then she went to the edge of the water and crawled under a grapevine for camouflage. All night she waited quietly for Jijogweh, but she saw not one creature stir.

"The evil one is busy somewhere else," Brave Girl thought, and she decided to return home and try again the next night. As she was gathering her things to leave, suddenly there was a loud, shrill, and terrifying shriek. And there, circling ominously over her on his giant wings was Jijogweh. She had expected him to be huge, but she was still struck with awe at his vast size.

She quickly drew her bow and reached for an arrow. But it hung soft and limp like a wet blade of grass, softened by the wet, night air.

But Brave Girl would not give up easily, and she knew she had something even more powerful than the arrows. She grasped the sacred pouch and called out, "Good spirits of the night, be my friends!" Jijogweh loomed closer, but she kept saying the words over and over.

She reached again for an arrow and now drew out a stiff, straight arrow that lay solidly in her bow. Flying from her ash bow, the arrow pierced the wicked heart of the bloodthirsty bird.

The monster fell, its wings beating the air wildly and thrashing the water until it foamed. Then as the evil bird sank beneath the surface forever, a flock of gulls rose like a cloud from the water, hovered a moment over the spot were Jijogweh had perished, and flew away.

The gulls had been eaten by Jijogweh and were released when he died. Now, as their way of thanking Brave Girl for freeing them, whenever a storm comes, the gulls fly up to warn the people.

Moisture Folklore

Here's another installment in our continuing series of classic weather lore that can be tied to verifiable weather facts. People long ago adopted short, poetic ditties that can sometimes accurately predict the weather. Here is an example:

> *Fog on hills,*
> *More water for the mills.*

Fog on hills is really low-hanging clouds that form as warm, moist air blows against the slope of a hill or mountain and is forced upward. As the air rises, it expands and cools, and the moisture in the air condenses into droplets. If the fog does not evaporate, it may later cause rain. Found generally in Virginia, North Carolina, and in mountainous sections of New England, and in Scotland, this poem is only a moderately reliable weather predictor:

> *When sheep gather in a huddle,*
> *Tomorrow we'll have a puddle.*

Sheep are sensitive to small, telltale changes in the weather because their wool traps air as insulation. Sheep huddle to keep warm and so may foretell the coming of cool air and rain. Found in the South and in New England, this one is a good weather indicator:

> *The owls hoot, peacocks toot,*
> *The ducks squawk, frogs yak—*
> *   twill rain.*
> *The loons call, swallows fall,*
> *Chickens hover, ground hogs take cover—*
> *   twill rain.*

Most animals are more sensitive than we are to changes in the pressure, temperature, and humidity of the air. Some changes in the weather, such as the low pressure and increased humidity before a storm, may make animals uncomfortable, restless, and noisy. Their behavior tells us that the weather is soon to change. Found in Virginia, Kentucky, England, and Scotland, this rhyme is a reliable indicator that weather will change.

Precipitation and Humidity Products

Sling Psychrometer
Robert E. White Instruments, Inc.
34 Commercial Wharf
Boston, MA 02110
(800) 992–3045

In the 1990s we've got digital equipment designed for just about anything you can imagine, and humidity measurement is no exception. Still, if you prefer the simplicity of traditional weather instruments, you may appreciate an industry-standard sling psychrometer. Since sling psychrometers are getting harder to find, you may want one simply as a potential collector's item.

White Instruments has two models. The traditional instrument used regularly by the National Weather Service is a full-size (12 inches long) model, which costs only $64. It has a stainless-steel backplate, chain links, and a turned wooden handle. It operates from –20° F to 120° F. Humidity conversion tables are included.

If you prefer compactness over tradition, you will want the Bacharach Pocket Sling Psychrometer. Still priced inexpensively and providing good accuracy, this model is but 8 inches long and is meant to be carried in its protective case. It includes an extra wick, coiled up inside, plus instructions and an integral psychometric slide rule. It operates within the same temperature range as the NWS model and costs $60.

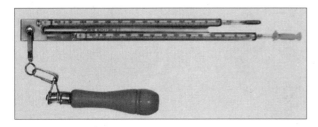

Nimbus Precipitation and Relative Humidity
Sensor Instruments Company, Inc.
41 Terrill Park Drive
Concord, NH 03301
(800) 633–1033
In New Hampshire: (603) 224–0167

Sensor offers two professional monitors that track precipitation and relative humidity. Both units are powered by four "C" cell batteries, have thirty-five-day memories for recording readings, which can be recalled later or downloaded directly into a PC, have optional PC interfaces, come in your choice of cherry or oak cases, and have three-year warranties.

The Relative Humidity instrument includes a sensor with a 50-foot cable so that you can easily check outdoor humidity. This monitor has a list price of $450. The Nimbus Precipitation monitor requires an automatic electronic sensor to feed it rainfall data and has a list price of $300. (See the listings below for compatible rain collectors.) The PC interface—each unit needs its own—lists for $100.

Rain Collectors
Robert E. White Instruments, Inc.
34 Commercial Wharf
Boston, MA 02110
(800) 992–3045

Don't call the RainWise Model 100 a rain gauge. It's a rain collector, and one of the best in the industry. It features an 8¾-inch-diameter stainless-steel bucket with an automatic tipping mechanism, which eliminates the need for manual dumping. Powered by a 9-volt battery, this professional-quality unit can be connected to standard weather-monitoring devices such as the Nimbus Precipitation unit listed above, a personal computer, and most home weather stations. It's highly accurate in measuring rainfall up to 99.99 inches. You don't need an expensive weather monitor, because its list price of $79.95 includes a resettable digital counter and a 30-foot connecting cable for remote mounting.

Bedside "Comfort Clock"
Wind & Weather
The Albion Street Water Tower
P.O. Box 2320
Mendocino, CA 95460
(800) 922-9463

Humidity and temperature have a great effect on your physical comfort. Find your comfort zone with this stylish quartz digital clock from Wind & Weather that combines a thermometer and hygrometer. It has a snooze alarm and displays time, a monthly calendar, and the year in a large, multifunction LCD panel. It can even call up past and future calendars from 1901 through 2099. It is priced at $75 plus $1.50 shipping and handling.

Professional Instrument Probes
Rotronic
160 East Main Street
Huntington, NY 11743
(800) 628-7101
fax: (516) 427-3902

Here's a source for high-performance, standard-setting humidity- and temperature-measuring instruments. Rotronic offers two professional meteorological probes. The first is the MP-100, which is designed primarily for use at remote locations where a power source is a problem. The MP-100's low power consumption and wide range of tolerable voltage makes it ideal for these sites. The TM-12 is designed for locations where a steady source of reliable power is available, such as on a utility pole. Both units feature drift-free outputs with temperature compensation and are highly resistant to contaminants and industrial pollution.

Amateur Rain Gauge
American Weather Observer
401 Whitney Blvd.
Belvidere, IL 61008

Made of durable, high-impact plastic, this inexpensive rain gauge is sold by the newspaper that is the voice of the Association of Weather Observers (AWO). Priced at only $26.95 (including tax and shipping), this is a bargain in a rain gauge that offers accuracy to 0.01 inch with an 11-inch capacity. It includes a rainfall log and mounting hardware. Organizations affiliated with the AWO qualify for quantity discounts.

Dispenza Rain Gauge
Copious Corporation
2758 Sailer Avenue
Ventura, CA 93001
(805) 644-0622

Copious claims this is the only commercially available rain gauge that can measure both snowfall and rainfall. It measures rainfall to 0.01 inch with 98.7 percent accuracy even in heaviest rainfall. The Dispenza requires no outside energy to operate, yet it produces a pulse that can be read by weather stations and adapted to electronic transmission, telephone lines, and chart recorders. Its patented automatic siphon valve lets it operate completely unattended. The price of $398 includes the gauge, a digital electronic counter, and 50 feet of cable.

Recording Hygrothermographs
Robert E. White Instruments, Inc.
34 Commercial Wharf
Boston, MA 02110
(800) 992-3045

White Instruments has a variety of hygrothermographs. Here are two that will serve most users.
Museums exhibiting prized treasures must con-

stantly monitor humidity and temperature to protect their priceless stores. The Belfort #5-594 Hygrothermograph is the industry standard and attains its superb accuracy using a gold-plated Bourdon tube temperature sensor and a hair-humidity sensor. It's available in two portable models. One is battery-powered, while the other is powered by a windup key. Both models track humidity and temperature for up to seven days and include a supply of charts, pens, ink, and an instruction book. Call White for pricing of this premier, professional hygrothermograph.

Another professional-quality hygrothermograph is the Maxant #173. Its performance is similar to the top-of-the line Belfort. It runs on an "AA" battery for portability. The Maxant includes all standard accessories but has a locking case to maintain data security. It's priced at $600.

Instrument Shelters
Wind & Weather
The Albion Street Water Tower
P.O. Box 2320
Mendocino, CA 95460
(800) 922–9463

Quality instruments need a good shelter to protect them to ensure longest life and highest accuracy.

Beginners may find that a small shelter mounted on a north wall or a post will serve their needs. The small model holds up to three instruments, has inside dimensions of 16 by 8 by 5 inches, and costs $180 plus $17 shipping and handling.

You can build a complete home weather station with the medium shelter, which comes with its own stand. The medium model holds a recording instrument and several smaller instruments. Its inside dimensions are 19¾ by 19 by 14¼ inches. It is priced at $580 plus $35 shipping and handling.

Wind & Weather also has large shelters for professional installations. Call for details and pricing. All models are made of clear pine, covered with three coats of white latex paint, fitted with louvers for ventilation, and include locks and keys.

Chapter 8
Clouds

A bright blue summer sky is dotted with puffy white clouds dancing about and constantly changing their shapes and sizes. Every once in a while, new clouds arrive out of nowhere and others vanish. And on some days the sun is hidden by a dense sheet of clouds. This ever-changing fresco over our heads gives us our most obvious clues about the weather. Clouds can be beautiful. Clouds can be ugly. Clouds can be scary. But they always are our guide to the weather.

Observing cloud formations in that great painting in the sky will enable you to make short-term weather predictions. There are so many variations in formations that few people can recall what each means, but we've found a lot of good reference material that can help. Check the product listings at the end of this chapter for charts, reference cards, and books that link cloud formations to an understanding of the weather they accompany. Anyone who can match clouds overhead with color pho-

Updraft Region of a Severe Thunderstorm

tographs from a good reference guide can make relatively accurate predictions of tomorrow's weather.

Perhaps the best part about using cloud formations to predict weather is that you will not need to buy or maintain expensive equipment. Nor will you need any technical weather knowledge. Cloud formations are the perfect traveling weather station—all you need is one compact, easy-to-carry reference guide and an eye on the sky.

Even without a photographic guide to cloud formations, you can learn general principles that can help you predict general weather patterns. In Chapter 9, "Marine Weather," we'll give you a comprehensive listing of cloud formations and wind directions that mariners use to predict weather. Of course, at sea mariners depend on their weather knowledge for survival. Landlubbers, however, may only need it to help plan an afternoon picnic. Here's a brief, easy-to-remember summary that will give you a basic idea of overall weather patterns represented by cloud formations and wind direction:

1. First check the wind direction by wetting a finger, holding it up high, and feeling which side turns cool. Increased evaporation on the side facing the wind will make it cooler. You may also be able to observe a flag or smoke to help you determine wind direction.
2. Next, use these guidelines:
   - Winds from the northwest, west, and southwest generally bring good weather. If it's raining in the morning with northeast winds, a shift to the west means the clouds will clear soon.
   - Winds from the northeast, east, and south generally bring bad weather. If the sky is cloudy and the winds take an easterly shift, expect rain soon.
   - If you see no clouds but the winds are shifting between southeast and southwest, you can soon expect to see clouds and rain, perhaps with strong winds.
   - Early-morning fog—clouds on the ground—or dew or frost on the ground generally means no rain that day.

You can see that observing cloud formations alone is not enough for totally accurate weather predictions. If there is no wind, then whatever weather you are experiencing will not change. It's the wind's movement of the clouds that changes your local weather. Thus, you must look at how clouds form.

How Clouds Form

Clouds are made of the same ingredients as fog, but they form high above the ground. Clouds need water just to be seen; they are formed by liquid droplets of water condensed from the vapor contained in rising currents of air. It's amazing how much water a cloud can contain. A typical rain cloud can weigh billions of pounds and contain trillions of water droplets.

Air grows cooler as it rises, with the temperature dropping 3.5° F for every 1,000 feet that the air rises. The invisible water vapor begins condensing into water droplets to form a cloud when the rising air cools to its dew point. A cloud will form only if the air above the rising air is cool. If the air is warm, it stops the lower air from rising.

Clouds appear when the rising air reaches the condensation level. Water molecules condense on the billions of minute particles blowing around—from grains of salt from the sea to dust and pollen from the land—and the clouds appear. The height of the condensation level depends on the amount of moisture in the rising air. This can vary from a few hundred feet to as much as 6 miles or more before clouds form. Some clouds are made totally of ice, because the moist air has risen so high (a cumulonimbus can reach as high as 60,000 feet) that the water molecules form ice crystals.

Whatever their makeup, clouds appear white to us. Since water is colorless and the sky appears blue to us, why is it that clouds look white?

Why Clouds Appear White

Bright white light is actually a blend of every color in the spectrum. The brightest white light in nature, the sun's light, contains every color, too. When the sun's light hits a droplet of water, the prism effect breaks the light up into individual colors that we then can see as violet, indigo, blue, green, yellow, orange, and red. Some of this colored light reflects against the back side of the droplet and bounces back out of the droplet.

The sky's familiar blue color results from the colors produced when tiny dust and moisture particles in the air scatter the sun's rays. Reflected light from the shorter side of the wavelength spectrum (blues and violets) scatter more widely than those from the other end of the spectrum (yellow and red).

As the number of moisture particles in the sky increases, more of the separated colors of the scattered light are reflected, not only the blue rays. That's when we see the sky taking on a whitish haze. When clouds form even more light is scattered, and we begin to see a recombined mixture of all the colors of the spectrum; thus, our eyes perceive white.

During sunsets and sunrises the sun is closer to the horizon, so we see its light through an angle closer to the Earth's surface. Thus, the sunlight must travel a longer path before reaching our eyes. This results in it striking more dust and moisture particles and scattering the longer wavelengths (yellows and reds) more widely, producing sometimes brilliant color displays in the West.

Cloud Classifications

Clouds come in many shapes and sizes, and different kinds occur with different types of weather. Identifying and interpreting clouds can aid in understanding and forecasting weather.

Clouds are classified according to how they are formed. The three basic types are heaped, layered, and feathery. Formally, these are called:

- *Cumulus clouds,* meaning piled up or accumulated, are formed by rising air currents. These clouds appear puffy and build upward. They are taller than they are wide.
- *Stratus clouds,* meaning sheet-like or layered, are formed when a layer of air is cooled below the saturation point without vertical movement.
- *Feathered clouds,* meaning wispy streaks.

There also is an international cloud classification that provides a standardized order for professional meteorologists. In this system clouds are divided into four families—*high clouds, middle clouds, low clouds,* and *vertical clouds*—and are then further classified by their shape and appearance.

In 1803 an English pharmacist named Luke Howard identified ten distinct categories of clouds based on variations of the four basic cloud forms. No one has yet been able to improve on this system.

A description of the major cloud types follows and should help to better understand their relationship to the weather. Howard's cloud-naming system may seem complex at first, but you soon will see that cloud names describe their type and form. A cloud that typically produces heavy precipitation has the word *nimbus* attached to it. Windblown clouds that are broken have the prefix *fracto-*. *Alto,* meaning "high," precedes the name of stratus or cumulus types of middle-layer high clouds.

1. High clouds are broken into three types, are made up almost solely of minute ice crystals, and their bases are at an average of 20,000 feet above the earth. These types are:

   - *Cirrus:* Composed mostly of ice crystals 5 to 8 miles up, where the temperature is far below freezing, they appear as feathery, thin, wispy strands.
   - *Cirrocumulus:* Formed at 20,000 to 25,000 feet, these thin, patchy clouds are infrequently seen. They are often rippled, forming a wave-like pattern.
   - *Cirrostratus:* Also formed at 20,000 to 25,000 feet, these thin, sheet-like clouds look like a fine veil. Cirrostratus clouds sometimes form a "halo" around the sun or moon because they are made of ice crystals.

Some Cloud Types

Cirrus

Cirrocumulus

Cirrostratus

Altocumulus

Stratocumulus

Cumulonimbus

2. Middle clouds are usually cumulus or stratus, and their bases can be found at about 10,000 feet above the earth. These two types are:

   - *Altocumulus:* Made of water droplets, these patches of puffy, gray, or white clouds have many variations in appearance and formation. Altocumulus clouds rarely produce precipitation. They often resemble cirrocumulus clouds.
   - *Altostratus:* Composed mostly of dense sheets of gray or blue, these can vary in thickness from very thin to several thousand feet. The presence of altostratus clouds in the area usually means that unfavorable weather and precipitation are nearby.

3. Low clouds range from near the Earth's surface to about 6,500 feet above the terrain, although the heights of these clouds may change rapidly. The three types of these clouds are:

   - *Stratus:* With the base above the ground, this low, sheet-like fog can make the sky appear heavy. Stratus clouds can produce only a fine drizzle because there is no vertical movement in them.
   - *Nimbostratus:* These true rain clouds often have streaks of rain stretching to the ground, and they appear darker than regular stratus clouds.
   - *Stratocumulus*: Although they do not produce rain, these dark-gray, layered masses of puffy clouds often fuse with the nimbostratus clouds, leaving the Earth wet with rain.

4. Vertical clouds of the cumulus type result from strong vertical currents. They can form at almost any altitude and may reach up to 14,000 feet. There are two types of vertical clouds:

   - *Cumulus:* Formed by day in rising air, these puffy, cauliflower-like clouds constantly change shape. Cumulus clouds mean fair weather unless they take on added vigor and pile up into cumulonimbus clouds.
   - *Cumulonimbus:* Known as *thunderheads,* the bases of these clouds nearly touch the ground.

When they become very violent, these clouds can produce tornadoes. During violent updrafts the tops of these clouds may reach 75,000 feet and spawn some of our wildest weather.

Myths (and Truths) about Clouds from Folklore

In chapter 2, "Weather Lore," we learned a lot about weather folklore. Most of it is pure fun, if not funny. Some weather folklore is roughly based on fact, but is hardly useful for predicting weather. Other folklore, however, is quite accurate and useful.

The most reliable of all weather folklore is that involving the observation of the sky and clouds. The shape and movement of clouds often are an accurate indicator of weather to come. Here are a few examples of such folklore:

- Red sky at night, sailor's delight;
  Red sky in morning, sailors take warning.
- Mists in the old moon, rain in the new;
  Rain in the old moon, mists in the new.
- When clouds appear like rocks and towers,
  The Earth's refreshed with frequent showers.
- If cumulus clouds are smaller at sunset than at noon, expect fair weather.
- Cumulus clouds in a clear blue sky, it will likely rain.
- The higher the clouds, the fairer the weather.
- Rainbow in the eastern sky, the morrow will be dry;
  Rainbow in the west that gleams, rain falls in streams.
- A ring around the sun or moon,
  Means that rain will come real soon.
- If you see the sun set in a cloud, it will rain tomorrow.

The sayings about the meaning of red skies at dusk and dawn rest on the principle that weather usually moves from west to east. When the western horizon is clear, the sun reflects from the clouds and fair skies will move in. But a red sky at dawn means that the sun is reflecting from high clouds moving in from the west, and these clouds often appear before a storm.

The refraction of sunlight by rain creates a rainbow. A rainbow in the west means rain is coming, while a rainbow in the east means the storm has passed.

A halo around the sun or moon—a common sign that rain will follow—is caused by light from the sun or moon passing through ice crystals in the upper stratosphere. These ice crystals can produce high cirrus clouds that meld together. So watching the sky and tracking its swiftly changing cloud cover can help you stay on top of the weather to come.

Rainbow

Native Americans across the continent had many different names for rainbows: "The Rain's Hat," "The Great Spirit's Fishing Line," "Strong-Medicine-to-Drive-Away-Rain," "Cloud-Boy's Sky Path."

The Zia regarded rainbows as paths that only war heroes were allowed to travel. But the Yuchi said the gods had stretched a rainbow across the sky to prevent any more rain from falling.

The Hoh and Quileute live on the Olympic Peninsula in the state of Washington. Here's a story told among their tribes about a mysterious rainbow woman who lived on the other side of the ocean.

One morning, a very long time ago, a hunter shot a duck with a newly made arrow. The arrow did not harm the duck but settled deep into her feathers. When she flew away, the arrow flew with her.

The hunter didn't want to lose his newly made arrow, so he followed the duck, hoping to get it back. He chased the duck all day and became completely lost on the other side of the ocean. When he reached the opposite shore, he saw the duck, but it had changed into a giant woman!

The giant woman grabbed the hunter and screamed, "Now I have you! I've been waiting to catch a new husband, and here you are!" She headed for her house, dragging the hunter alongside by his long hair.

Then they passed one of her neighbors, who whispered a few words to caution to the hunter, "Watch out! Tonight she'll kill you and eat you, just like the others."

Late that night when the hunter thought the giant was asleep, he escaped, dashing through the trees, over the hills, and along the water until once more he was lost. He stopped running, but then he heard the giant woman's thundering footsteps bearing down upon him.

Already exhausted, the hunter knew he couldn't outrun the giant, so he sat down and waited. And then a beautiful woman appeared next to him. She was dressed in a cloth of many colors. The hunter asked, "Who are you?"

"I am Rainbow Woman," said the giant. "Why are you running so fast?"

"Can't you feel the earth tremble?" said the hunter. "Those are the footsteps of an evil, giant woman who wants to kill me."

"Then I can help you," replied Rainbow Woman. "Run on ahead. I'll follow in a moment."

When Rainbow Woman caught up with the hunter, they could hear the giant woman thumping down the path, shaking the trees. Then there was one loud crash, and everything was silent.

"What did you do?" asked the hunter.

"I set a trap," said Rainbow Woman. "Now, come along and rest at my house," she said, for already she loved him.

So the hunter went off with the beautiful Rainbow Woman, and by morning he had fallen in love, too.

The hunter married Rainbow Woman and after a while, they had a child. But one morning the hunter took his bow and arrows, as he often did, and went hunting. When he had not returned by late afternoon, Rainbow Woman went to look for him. But she could not find him. To this day, he has not returned.

Rainbow Woman knows her husband has just lost his way again, so she keeps watch faithfully. Sometimes she climbs into the sky so she can see better to search the land. When you see a double rainbow, you know that their child is helping her mother search for the lost hunter.

The SkyWarn System

Once you've learned how to interpret the weather by looking at cloud formations, you are likely to find your eye on the sky more than ever before. If so, don't let this newly trained weather eye of yours go to waste—use it to help the National Weather Service's SkyWarn system. Participating in SkyWarn will encourage you to use and develop your cloud-watching skills.

SkyWarn is a national system of volunteer amateur weather observers who are specially trained and certified. SkyWarn weather watchers take most of their clues from the cloud formations they see. Observations from many SkyWarn participants can be pieced together by the National Weather Service to create a responsive early-warning system for hazardous weather. The SkyWarn program saves thousands of lives every year.

SkyWarn is an integral part of the total weather-forecasting system of the National Weather Service. While sophisticated, advanced systems like NEXRAD will locate and track a potentially severe and dangerous storm, SkyWarn spotters help the NWS determine what that storm is actually producing, whether it be a tornado, hail, strong winds, or heavy rains. NWS forecasters then use all available information to issue appropriate warnings for areas that may be impacted by the storm.

To join the SkyWarn network, contact the NWS at:

Warning and Preparedness Meteorologist
National Weather Service Forecast Office
RD #1, Box 107
Sterling, VA 22170
(703) 260–0209

The NWS will mail you an application. After you are registered you will be scheduled for SkyWarn spotter training in your local area. After your training session, you will receive a SkyWarn-spotter ID card and registration number along with a packet of spotter information. The National Weather Service will file your mailing information and add your location to the SkyWarn map so it will know how to locate you to request information from time to time. The NWS mails periodic updates, activities calendars, information, and notices of optional advanced training to registered SkyWarn spotters.

SkyWarn Reporting Procedures

1. Contact the National Weather Service:
   A. Amateur radio
   B. Telephone (toll-free 800–253–7091, or 703–260–0206)

2. Identify yourself:
   A. "SkyWarn spotter"
   B. SkyWarn ID code

3. Location of the event:
   A. County and state
   B. Position relative to known town/ landmark
   C. Your position relative to the event/ storm

4. Event (type of severe weather being reported):
   A. Tornado (on ground); funnel (not touching ground)
   B. Storm rotation; rotating wall cloud
   C. Hail—size and depth on ground (if applicable)
   D. Wind—50 mph or greater; indicate gusts or sustained
   E. Heavy rain—1 inch or more (give time duration) or any indication of flooding occurring or about to occur
   F. Damage—wind, tornado, hail, flood; if trees or branches, try to estimate size and type (pine, hardwood, etc.)
   G. Ice accumulation—1/8 inch +; surfaces (roads, trees, etc.)
   H. Snow accumulation—4 inches +; road versus grassy areas
5. Time:
   A. When did the event/storm occur?
   B. Duration, begin/end time?
   C. Is it still occurring?

Doing Something about the Weather

People have long yearned to do something about the weather rather than idly sitting by and accepting whatever it hands down to us. Although little can be done, we have developed a particular fascination with trying to "make rain."

Rain is so important to our survival that desperate people will believe even fraudulent tricksters if there is any chance that they might help them survive. Over the centuries countless charlatans have claimed to be "rainmakers," duping farmers with rosy promises of crop-saving rains. Indeed, few rainmakers were more than frauds. Perhaps their greatest skill was a keen understanding of how to read clouds and predict the weather—in times when weather was poorly understood—and they went to drought-stricken areas with such perfected timing that they appeared to create rain, which already was on its way.

Among Native Americans, the rain dance is among the most sacred and important rituals. Any holy man who was thought to have the "medicine" to make rain was highly honored in his tribe. Again, these wise men may have had no more than a natural inclination to understand weather by linking cloud formations that they had observed with the weather that followed.

Today's rainmakers are neither charlatans nor holy men. Instead, they use well-known basic weather principles to produce clouds that will in turn produce rain. Their work, known as *cloud seeding*, is neither trickery nor very spiritual, but it has proven to be effective.

Cloud seeding is a modern scientific approach to rainmaking. Using an airplane, today's rainmakers release dry ice, salt particles, water spray, or other matter into existing clouds. These particles "seed" the cloud, increasing its ability to produce rain. Some substances, such as silver oxide, are released on the ground and carried into the clouds by winds. Whatever material is used, water droplets in the cloud form around the particles, forming raindrops. Cloud seeding, therefore, does not create rain out of nothing. A considerable amount of moisture must be present first. The seeding merely pushes marginal clouds over the edge to get them to produce rain.

While you easily can imagine cloud seeding being of interest to drought victims, it has other, unexpected uses. Sometimes it is used in areas that have had too much rain to get approaching clouds to dump their rain elsewhere, thereby lessening the severity of storms where rain is not wanted.

Weather scientists are still experimenting with cloud seeding and other techniques, such as sending high electric current through clouds. Of course, the hope is that they will learn to control, or at least affect, the weather on a truly major scale, so that hurricanes can be diverted or dispersed. However, for all our high-tech hardware, supercomputers, satellites, and a vast store of knowledge, experts can do little more than occasionally increase the likelihood of local rains.

> ## Fog
>
> According to a Tsimshian story about fog, the daughter of the Great Chief in the Sky dips her long skirts into the waters of the sky country. When she wrings out her clothing over her father's fire, she makes a cooling fog that blocks the heat of the sun and refreshes the hot, tired people below.
>
> The Crees originally lived in the Canadian forests, but bands of them moved south onto the plains and became buffalo hunters. There they were known as the Plains Cree. They told a tale about fog that depicted it as being the spirit of evil, formed into an earthly vision.

Ultraviolet Warning Index

Ultraviolet light (UV) from the sun is responsible for sunburn. Ultraviolet rays can damage your skin, even through some cloud cover. Damage potential from UV varies according to a range of factors that include the season, the condition of the ozone layer, barometric pressure, temperature, elevation above sea level, and, of course, cloud cover. Damage from UV accumulates over time, with much of the injury done when people are youngsters. The Solar Warning Index forecasts the amount of dangerous ultraviolet light from the sun that will be arriving at the Earth's surface at noon.

Minimal (0–2)
   Very fair people may burn in 30 minutes, while those less susceptible to damage may be safe for up to two hours.

Low (3–4)
   Safe time for pale people is 15 minutes, while those with darker skin will suffer damage in 75 minutes.

Moderate (5–6)
   Fair people are safe for 10 minutes, up to 60 minutes for those with more protective skin types.

High (7–9)
   Safe time is only 7 minutes for unprotected people with fair skin, while people not susceptible to burning are safe for only about 35 minutes.

Very High (10+)
   Safe time is a mere 4 minutes for those most at risk, and even those people whose skin is least affected by the sun can suffer damage after 20 minutes of unprotected exposure.

Don't take chances, whenever there is a Solar Warning Index hazard, use sunscreens, sunglasses and minimize your exposure to direct sunlight.

—Courtesy of Stormfax® Weather Services

Cloud-Related Products

Cloud Charts and Book
Cloud Chart, Inc.
P.O. Box 21298
Charleston, SC 29413–1298
(803) 577–5268

Cloud photos are Cloud Chart's business, and its specialties are wall charts and laminated 11-by-15-inch cards. The charts show various cloud formations around the world, coded and indexed with official World Meteorological Organization names and numbers. Each photo on the chart includes a caption describing the cloud and offering tips on interpreting forecasts based on the formation. These make excellent teaching tools, as the chart groups are written for different grade levels, from elementary through high school.

Cloud Charts also offers *The Weather Wizards Cloud Book,* originally written by the late Louis D. Rubin, Sr., but now updated by Jim Duncan, a graduate meteorologist and television weather anchor in Richmond, Virginia. The new book features a stunning array of cloud photographs in full color, keyed to the text so you can use them to predict weather. It also covers how weather conditions change, cold and warm fronts, thunderstorms, hurricanes, tornadoes, sky colors, folklore sayings, curious truths and untruths about the weather, volcanoes, and the weather, and gives instructions on building a home weather station.

Sky Guide
Sky Guide
P.O. Box 30027
Greenwood Station
Seattle, WA 98103
(206) 782–8485

The Sky Guide names and describes all major cloud types and identifies them on a poster with photographs taken around the world. Sized at 25 by 38 inches, this poster is the largest cloud chart we've found. It contains thirty 4-by-4-inch color cloud photographs and includes a set of thirty matching color slides. The price is $7.95, plus $3.00 for shipping and handling.

Cloud Chamber Kit
Hubbard Scientific, Inc.
P.O. Box 760
Chippewa Falls, WI 54729–0760
(800) 323–8368

Hubbard offers a teaching kit that allows you to see and trace radioactive elements. You'll be amazed as it reveals trails of radioactive rays. This kit includes a plastic cloud chamber, a radioactive source (not harmful, of course), and the magnet you'll need for the experiments. It also includes a comprehensive teacher's guide that explains clouds, and how to use the kit in experiments. Single kits are $6.35, but Hubbard offers a special teacher's pack of ten kits for only $29.95.

Satellite Cloud Photographs
OFS WeatherFax
6404 Lakerest Court
Raleigh, NC 27612
(919) 847–4545

OFS WeatherFax II is an advanced hardware and software package that enables you to capture and display high-quality APT weather-satellite imagery on an IBM-compatible personal computer. When attached to the audio output of your receiver, it captures pictures directly from orbiting satellites, providing immediate, high-resolution cloud-cover images. You then may enhance these cloud images using the extensive tools included in the package. Images can be printed to nearly any graphics printer and saved in standard TIFF or GIF formats that can be used by many other weather software applications.

Lightning Sensor
McCallie Manufacturing Corporation
P.O. Box 17721
Huntsville, AL 35810
(205) 859–8729

CHAPTER 8: CLOUDS

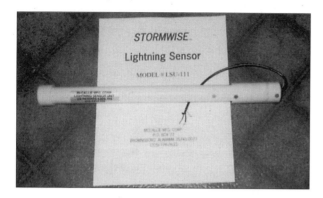

Clouds can hide lightning and keep you from spotting the buildup of a severe storm by eyesight alone. Invisible intracloud lightning starts ten to thirty minutes before a severe storm produces the startling, visible lightning that we can see. A lightning sensor can help your sky-watching considerably, giving you advance warning of approaching or developing storms. The SAU-444 "Atmosferics" Sensor by Stormwise is a good choice and includes a thirty-five-page instruction booklet so you'll know what its output means to you. The book also tells you how to build a warning system incorporating the sensor, using a few inexpensive electronic parts available in local stores. With the unit mounted on a pole and the electronics connected, you can adjust its alarm to warn you of storms between 5 and 200 miles away. Powered by a 9-volt battery, the SAU-444 costs $46 plus $5.00 shipping and handling.

The Cloudless Earth—Sculpted in Glass

The Nature Company
Catalog Division
P. O. Box 188
Florence, KY 41022
(800) 227-1114

We've shown you how to find plenty of photos of clouds over the Earth, but here is a work of art that enables you to see the Earth as it really looks underneath its constantly changing robe of cloud cover. In a virtuoso performance of glassblowing art, the famed Lundberg Glassblowing Studios has created a masterpiece, 3½-inch-diameter glass globe based on a composite of satellite photos. In amazing detail you'll be able to see all the land features, such as coastal contours and island chains and vegetation. A truly unusual view of our planet, complete with crystal-clear cradle. The price is $110.

Chapter 9
Marine Weather

Mariners survive by their weather knowledge. Being in such intimate contact with the power of the waves, the wind, and the weather, they know better than most the extent to which the weather deserves respect, even today.

Today's high-tech world of power-driven vessels equipped with radio, radar, electronic marvels, and satellite communications can breed complacency. And most ships have redundant systems, especially those capable of long voyages. Yet the practical value to mariners of having a strong weather knowledge is as high as ever. The sea and the weather have not become less powerful; instead, we have merely become more adept at predicting what they might do next.

Some oceanographic experts say that our scientific, high-tech, competitive age has increased the level of maritime hazards. Why? Because with today's equipment, shipping-company managements tend to believe that the ships they own can go anywhere, anytime, in any weather. After all, they've paid small fortunes to outfit their vessels with an incredible array of expensive, up-to-date technology. So captains put to sea today with weather forecasts that would have kept them tied to the pier a few decades ago.

The oceans are still vast, even in relation to the speed of today's ships. Weather that is clear when a voyage begins can easily become hazardous at sea. The major cause of casualties, losses, and damage at sea remains the weather and the sea conditions it brings. Even professional oceangoing ship captains need constantly to cross-check the electronic data they receive and ensure that it makes sense based on what they see in the sky ahead.

Of course, if you're a captain of something a bit smaller than several hundred thousand tons of steel and you navigate on something a little more sedate than the stormy Atlantic Ocean, you may not have the best of today's remote weather equipment. Even if you do you are unlikely to have a complete redundancy. You could lose a crucial piece of equipment or perhaps experience total electrical failure. So few things will aid your safety on the water like a solid base of old-fashioned weather knowledge that doesn't require electrical power or someone else maintaining persnickety, complex equipment.

General Maritime Weather Rules

Remember that sailors circled the globe for thousands of years without today's high-tech weather equipment. To be truly safe, boating demands weather expertise, but acquiring it will require much study and the reading of many weather books. Even meteorology experts are constantly learning more about the weather. The study never ends. The good news, though, is that you can still sail safely using the less sophisticated equipment and knowledge handed down from centuries at sea.

The key to using hand-me-down maritime weather lore is to cross-check it and look for more than one sign. More than 2,300 years ago a Greek poet named Aratus advised weather caution by saying, "Make light of none of the warnings, but it is a good rule to look for sign confirming sign. When two point the same way, forecast with hope. When three point the same way, forecast with confidence." Carry the following listing of weather signs with you and watch for two or more signs that confirm one another:

Typical Marine Barometer—Essential to Sailors

1. Expect bad weather to clear when:
   - the barometer rises rapidly
   - cloud bases rise
   - winds shift to any westerly direction (especially if it shifts from east through south to west)
   - there's been a cold front passage within the last three to six hours

2. Expect good weather to remain when:
   - the barometer is steady or rising
   - you see a heavy dew or frost the night before
   - the sky is clear and the winds are light the night before
   - clouds are moving out and/or their bases are rising
   - the evening sky is clear and the setting sun looks like a ball of fire
   - the wind is steady and gentle from the west or the northwest
   - summer fog burns off in the morning
   - temperature is normal

3. Expect the weather to deteriorate when:
   - the barometer falls steadily
   - there is heavy rain the night before
   - temperature is much higher or lower than normal
   - temperature rises rapidly in the winter
   - the wind shifts to the south or the east (especially if it shifts from north through east to south)
   - clouds move in different directions
   - cirrus clouds become cirrostratus, their bases descend, and they increase
   - you see dark altocumulus or altostratus clouds in the west and the barometer is falling
   - there is a strong wind early in the morning

4. Expect rain or snow when:
   - the barometer falls steadily
   - you've noticed cirrus clouds thickening and lowering within the past day or two
   - the morning temperature is unusually high, humidity is high, and you see cumulus clouds building
   - the western sky is dark and threatening
   - a north wind "backs" (shifts in a counterclockwise direction from north to west to south)
   - a south wind increases in speed and the clouds are moving west
   - a warm or occluded front is approaching
   - you hear static on the radio and cumulus clouds are building (you can expect rain within an hour)

5. Expect fog when:
   - the sky is clear, the winds are light, and the air is humid the night before
   - warm rain is falling ahead of a warm front

CHAPTER 9: MARINE WEATHER

- water temperatures are warm and the air is much colder
- warm winds are blowing humid air across a much colder surface (either land or sea)

6. Expect temperatures to rise when:
   - a warm front has passed
   - westerly winds shift to the northwest or the south
   - the daytime sky is clear with moderate southerly winds
   - the nighttime sky is overcast with moderate southerly winds

7. Expect temperatures to fall when:
   - a cold front has passed
   - the barometer is rising steadily in the winter
   - winds are steady from the north or northwest
   - the sky is clear at night with light winds
   - the winds shift toward north or northwest

The Basics

The two main keys to predicting weather are wind direction and air pressure. The following table provides a summary of the basic rules of thumb of wind and pressure that will help you predict the weather with reasonable accuracy. Usable since Torricelli invented the barometer in 1644, they are still valuable aids despite today's sophisticated radar and satellite images via weather faxes.

Wind direction	Pressure (inches)	Prediction
SE to NE	30.10 to 30.20 and falling slowly	Rain within twelve to eighteen hours and increasing winds
	30.10 or below and falling rapidly	Rain within twelve hours and increasing winds
	30.00 or below and falling slowly	Rain will continue one to three days (or longer)
	30.00 or below and falling rapidly	Rain soon, with high winds, clearing in thirty-six hours
SW to NW	30.10 to 30.20 and steady	Fair, with little change for one to two days
	30.10 to 30.20 and rising rapidly	Fair, but rain within two days
	30.20 or above and steady	Fair, with little change
	30.20 or above and falling slowly	Fair for at least two days, warming
S to E	29.80 or below and falling rapidly	Severe storm in a few hours, clearing within twenty-four hours
S to SW	30.00 or below and rising slowly	Clearing in a few hours, remaining fair several days
S to SE	30.10 to 30.20 and falling slowly	Rain within twenty-four hours
	30.10 to 30.20 and falling rapidly	Rain within twelve to twenty-four hours, increasing winds

Wind direction	Pressure (inches)	Prediction
E to NE	30.10 or above and falling slowly	Fair with light winds for two to three days in summer; rain within twenty-four hours in winter
	30.10 or above and falling rapidly	Rain within twelve to twenty-four hours in summer; rain or snow within twelve hours with increasing winds in winter
E to N	29.80 or below and falling rapidly	Heavy rains and a severe storm within hours; snowstorm and colder in winter
Shifting to W	29.80 or below and rising rapidly	Clearing and colder

The Beaufort Scale

Professional mariners can tell wind speed by looking at the condition of the sea. Waves and swells serve as their anemometer. In 1805 British rear admiral Sir Francis Beaufort saw the need to set a standard so that sailors could easily communicate their wind observations and convert them into meaningful reports. That was in the days of sailing ships, however, and his original wind scale was written in terms of how the canvas on a tall ship would behave. For example, the highest wind on Rear Admiral Beaufort's scale, 17, was defined as "That which no canvas could withstand." Today we define the *Beaufort scale* in terms of actual wind speeds, measured 10 meters above the surface.

BEAUFORT SCALE

Beaufort Number	Name	mph	Knots	Effect at sea	Effect on land
0	Calm	0	0	Glassy sea	Smoke rises vertically
1	Light air	1–3	1–3	Scalelike ripples, no crests	Smoke drifts
2	Light breeze	4–7	4–6	Small wavelets, crests, but they don't break	Wind is felt on face, leaves rustle
3	Gentle breeze	8–12	7–10	Large wavelets, crests start to break	Light flags move, leaves move
4	Moderate breeze	13–18	11–16	Small waves, whitecaps begin to form	Dust and loose papers move, branches move
5	Fresh breeze	19–24	17–21	Moderate waves, many whitecaps	Small trees move
6	Strong breeze	25–31	22–27	Large waves, whitecaps, and some spray	Large branches move, umbrellas are hard to hold

Chapter 9: Marine Weather

Beaufort Number	Name	mph	Knots	Effect at sea	Effect on land
7	Moderate gale	32–38	28–33	Breaking waves with white foam, spindrift forms	Large trees move, walking is difficult
8	Fresh gale	39–46	34–40	Moderate high waves, break into spindrift, foam	Twigs break, walking is very difficult
9	Strong gale	47–54	41–47	High waves, dense foam streaks, sea rolls	Shingles blow off roofs
10	Whole gale	55–63	48–55	Very high waves with long crests, foam is blown in dense white streaks, heavy rolling white sea	Trees are uprooted, moderate damage to buildings
11	Storm	64–73	56–63	Extremely high waves, vessels seem to disappear into valley of swells, sea is covered in foam	Widespread, extensive damage
12–17	Hurricane	74–	64–	Air is filled with foam and spray, visibility bad	Violent destruction

Horse Latitudes

Though the Beaufort scale originally was developed for mariners, its usage has become popular on land as well. Some maritime wind terms, however, will never find their way ashore. A well-known example—now obsolete—is the term "horse latitudes."

Early international commerce depended on the wind to carry the ships of trade across the oceans. The only source of power for propulsion was the gift of the wind. This limitation could spell trouble in the climatic belt that lies between 25° N latitude and 30° S latitude, because of its light rainfall and light winds.

In the days when the tall ships ruled the seas of commerce, a freighter could become becalmed in this global band of light winds, meaning that the winds went dead, making their swift, proud ships move no more than paintings on a perfectly smooth, glassy seascape. Captains had no way of knowing how long their ships would be stuck and had to prepare for the worst.

To save the lives of the men in his crew, many a tall-ship captain ordered the horses jettisoned overboard so the load would be lighter and the food would last longer. It's easy to imagine that the sight was hard to forget. All around their ship were horses that moments ago were valuable cargo but that now bobbed up and down in a still, green sea, eventually to die. Even if a sailor never had seen horses jettisoned on his own ship, nearly all saw the floating carcasses of horses from other ships, sacrificed for the lives of the men on board.

When sailors crossed into this unpredictable, and sometimes dreaded, climatic band, they steeled themselves for what they might see in the horse latitudes.

The Weather Luck of Christopher Columbus

In August 1508, the Spanish explorer Ponce de Leon encountered two hurricanes. The first drove his ship onto the rocks in the port of Yuna, Hispaniola ("discovered" 16 years earlier by fellow explorer Christopher Columbus); 13 days later a second cyclone beached the same vessel on the southwest coast of Puerto Rico.

Many years later, another famous Spanish explorer, Hernando Cortes, whose discovery of treasure focused the covetous gaze of Europe on the newly discovered lands, lost the first vessel he sent to Mexico in a severe hurricane in October 1525. Captain Juan de Avalos (a relative of Cortes), two Franciscan friars and some 70 seamen were drowned.

Soon annual treasure fleets were carrying the riches of the virgin continent back to the war-depleted coffers of Spain. Each year the fleet assembled in Havana, scheduled to sail for Spain in March. Each year fiestas, banquets, and religious ceremonies stretched the departure date to August, and even to September, at the height of the hurricane season. Many ships sailed from Havana; often only a handful reached Spain. It is said that a ship was sunk for every lonely mile of the unexplored Florida coast.

Christopher Columbus made his first voyage to the New World in September and October 1492. These months mark the height of the North Atlantic hurricane season, yet Columbus did not encounter even nominally severe weather. Considering all the ships the Spanish lost over the decades after Columbus's early voyages—when the route to the New World was well-charted and the ships and navigation techniques they used were much improved from Columbus' day—how did he manage to make the voyage without losing ships and men to a hurricane? Was he truly a sailor of legendary skills, or merely lucky?

That question was answered in part in a study of the Atlantic crossing phase of the voyage, published in the February 1992 *Bulletin of the American Meteorological Society*. In that work, hurricane tracks for 1886-1989 were compared to the route taken by Columbus. The findings were quite surprising. Despite a frequency of about five hurricanes a year during September and October, only once in those 104 years did a hurricane approach within 2° (roughly 130 miles) of Columbus' track. It seems that the chances of Columbus hitting a hurricane on the open Atlantic were actually quite small—less than one in a hundred.

But what about the hurricane-haunted Caribbean? If we use Columbus' calendar (nine days behind ours), he sailed among several Caribbean islands for two weeks before eventually arriving off the coast of Cuba on October 28, 1492. Did he need "weather luck" to survive the island passages?

What we know about the weather Columbus encountered on his first voyage comes from a copy Fray Bartolome de las Casas made of Columbus' original *diario* or log. Most of the entries Columbus made while sailing the Caribbean deal with his impressions of the islands and their natives. A complete analysis of the Caribbean weather entries may be found in an article in the December 1991/January 1992 *Weatherwise*, entitled "Columbus Weathers the Bahamas."

Because the original Columbus *diario* has been lost, there is much debate about the accuracy of the Las Casas version. This is particularly true of Columbus' track in the Caribbean. No less than 15 tracks have been proposed by Columbus scholars. Possible landfalls range from Eleuthera (25° 05'N, 76° 10'W) to Turks Island (21° 40'N, 71° 45'W). At first glance, this uncertainty would appear to make it difficult to apply the same research technique that was used for the Atlantic crossing.

Hurricanes, however, usually cover such large areas that the differences between many of the hypothesized tracks are relatively unimportant; if a hurricane crossed almost any one of the routes researchers suggest Columbus sailed, it would undoubtedly have had a major impact on most of the other routes. Accordingly, a Caribbean box was constructed that contains most of the paths advocated by researchers. The dimensions of this box tend to favor the path suggested by Samuel Eliot Morison, one of the leading Columbus authorities, in his book *Admiral of the Ocean Sea* (1942).

Next, a computer search determined the number of hurricanes that crossed the box between the modern dates of October 20 and November 6 (corresponding to the old-style dates of October 11 and October 28) during the 104 years from 1886 to 1989. The computer search found 59 tropical cyclones (including both tropical storms and hurricanes) in the Atlantic Ocean and Caribbean during the period. With this information, the odds of a storm passage through the Caribbean box during Columbus' cruise among the islands of the New World could be determined.

As with the Atlantic crossing, Columbus could hardly have timed his Caribbean voyaging any better. During the 104 years of modern hurricane records, *not one* hurricane passed through the box during the dates when Columbus would most likely have been in the area. Seven tropical storms did cross the box during the period. Of these, four reached hurricane intensity during their lifespan, but not in the area where Columbus would have most likely sailed. To put the results in probability terms: The odds of encountering a hurricane while in the Caribbean box in the last half of October (or, using our current calendar, the last weeks of October and the first week of November) are less than once a century. The chances of encountering a tropical storm are higher, with a probability of 7:104 or 6.7%, less than once every 15 years.

So even in the storm-crossed Caribbean, the odds of Columbus encountering a hurricane similar to those of his Atlantic crossing were very low. Of course, that didn't guarantee he wouldn't hit one, so a final question remains—what might have happened if he had?

In fact, Columbus did encounter a hurricane on his fourth and final voyage to the New World. In 1502, he traveled from the Canary Islands west of Africa to Martinique in a swift 21 days, then sailed on to Santo Domingo. Arriving off the island of Hispaniola, the aging Admiral read the signs of an approaching storm that weather historian David Ludlum now classifies as a hurricane, and sent a captain ashore to warn [Nicholas de] Ovando, the governor. Ovando, however, ignored the warning and even refused to allow Columbus to seek shelter in Hispaniola's harbor. Instead, he ordered a large fleet loaded with gold and Indian slaves to set sail for Spain immediately. When the hurricane hit, it sank 20 ships and drowned some 500 men. Meanwhile, Columbus and his small fleet rode out the hurricane safely anchored in an island cove.

—By Randall S. Cervaney and Jay S. Hobgood. Reprinted with permission from *Weatherwise*.

Whatever his "weather luck," Columbus obviously was a weatherwise seaman and a very good bad-weather sailor.

Of course not all sailors have had the same luck. Seafaring lore is rich with stories of tremendous disasters caused by the weather. Most of those disasters, as you might expect, were caused by terrible storms that wrecked a ship, caused it to go astray, or simply swamped the ship under a wild and stormy sea. Can you imagine, though, that a case of extremely *good* weather could contribute to the sinking of a ship?

As unlikely as the theory in the next story may seem, some experts think that one famous sunken ship might have completed an Atlantic crossing uneventfully if the weather had not been so perfectly serene.

Weather and Oceanographic Charts

Columbus probably would have killed for the maritime information that bounces around in the atmosphere today. More than fifty transmitters all over the world broadcast weather and oceanographic charts of every description on a regular schedule, just waiting for you to flip the "On" switch on a receiver. These charts are enormously valuable to both pleasure and commercial craft by helping sailors enhance safety, improve passenger comfort, save time and fuel, and improve efficiency. For example, commercial fishermen can use the charts to locate fish since the movement of schools are dramatically affected by a variety of weather conditions. Sea temperatures and ocean-current charts, broadcast from many of the transmitters, will keep sailors up to date on those conditions.

Storms have a negative impact on equipment, passengers, and crew. Vessels that carry passengers can use wave-condition, wind, and ocean-current charts to plan the most efficient course. Avoiding unfavorable conditions not only saves fuel and time but also improves comfort and crew morale.

Broadcast surface analysis and prognosis charts show both existing and forecast positions of weather highs and lows as well as speed and direction of movement. Available on-line satellite photographs show cloud patterns and storm development. Equipment that will download these products includes information on broadcast schedules and frequencies for all worldwide stations. (See the product listings at the end of the chapter for compatible equipment.)

Marine Folklore

Here's the marine weather installment of our continuing series on weather poems that have some scientific basis in fact. This popular ditty, well known in New England and other coastal states, has proved to be a reliable weather forecaster:

> *When mist comes from the hill,*
> *Good weather it doth spill;*
> *When mist comes from the sea,*
> *Bad weather there will be.*

Mists or fogs on the tops of hills are usually shallow and do not have enough moisture to produce rain. They are easily dissipated by the sun's rays as the day progresses. Mists from lakes, broad rivers, and the oceans, however, contain much more water, making rain more likely. Many sea fogs are the result of the condensation that occurs when warm, moist air blows over water that is cooler than the air. When large quantities of warm, moist air blow over water that has a lower temperature, the resulting fog can be quite thick and hazardous.

CHAPTER 9: MARINE WEATHER

Weather and the *Titanic*

On its maiden voyage from Queenstown, Ireland, to New York City, the *Titanic*, the newest luxury liner afloat—and heralded as "unsinkable" due to recent technological improvements—carried 2,200 people, many of them immigrants bound for a new and better life in the New World. Instead, it went down on a calm, cold, starry night more than 80 years ago, on April 15, 1912.

At 20 minutes to midnight on that fateful night, the *Titanic* scraped its hull against an iceberg. Three hours later, the bottom of the North Atlantic became the final resting place for 1,500 people.

This tragic story still moves us 80 years later—the too few lifeboats, the unheeded iceberg warnings, the wireless operator of the nearby *Californian* retiring for the night, the *Californian*'s captain failing to respond to the *Titanic*'s distress flares.

The *Titanic* carried some of the wealthiest people of the time—and a few decks below, some of the poorest. The fateful night brought tales of great heroism and of base cowardice. In a matter of just three hours, all the unlikely elements of a great tragedy came together to make the story of the *Titanic* one of the most famous and revisited disasters of all time.

A Cold, Brutal Night

As a meteorologist, one thing has always stuck in my mind about the *Titanic* story—the weather that fateful night. Every eyewitness account vividly describes the night of April 14–15, 1912, the same way—the breathtaking cold, the incredible star-filled sky, the mirror-like state of the sea.

Putting aside the emotional imagery of the scene, the description of that night's weather—cold, crisp, calm, and clear—contains the classic features of what meteorologists refer to as an "Arctic high." An Arctic high is a large area of high pressure that originates in the Canadian Arctic and is associated with clear, cold, and calm conditions. Of course, it was much more than that to the people on the *Titanic*. For them, it meant a cold, brutal night that would forever haunt the memory of survivors.

Wanting to find out more about the weather on the *Titanic*'s ill-fated voyage and any role it may have played in the tragedy, I decided to consult the historical weather maps. Sure enough, on the map for April 15, 1912, there it was—almost eerie in its presence—a large 1037-mb Arctic high sitting almost directly over the site of the *Titanic* sinking. Deciding to investigate further, I checked the weather maps for the days preceding the disaster. With these, I was able to come up with a pretty good idea of the weather the *Titanic* probably encountered on its first and final voyage.

Thursday, April 11, 1912

The *Titanic* left Queenstown, Ireland, around 1:30 P.M. The 1:00 P.M. weather map showed the British Isles in a brisk northwesterly flow sandwiched between a low over Scandinavia and a high-pressure ridge over the Atlantic approaching Ireland. At the time of departure, the weather at Queenstown was generally cloudy, with a brisk north-to-northwest wind of 15–20 knots and an air temperature of about 50° F.

As the *Titanic* sailed westward into the high-pressure ridge that afternoon and evening, the clouds became scattered and the wind gradually diminished. That night, the *Titanic* sailed through the ridge with generally clear skies and light winds.

Friday, April 12, 1912

The ridge passed east of the *Titanic* overnight, and the winds backed to the southwest at about 15 knots. With a WSW bearing, the *Titanic* was encountering a nearly direct headwind. This would be true for most of its journey. The second morning of the voyage dawned with sunshine. By noon, the air temperature was around 60° F.

As the *Titanic* continued westward that afternoon, skies gradually clouded over as a weakening cold front approached from the west. According to the weather map, some scattered showers were reported northwest of the *Titanic*'s position, but no mention of the *Titanic* encountering showers has ever been found. If there was any shower activity, it must have been brief and light and had little impact on passengers or crew.

Saturday, April 13, 1912

The *Titanic* sailed into a fairly steady headwind of about 20 knots through the night under generally cloudy skies. Saturday morning dawned cloudy, though there were breaks in the cloud cover. Temperatures by noon were again around 60° F. The ship was gradually getting closer to a second cold front that lay to the west and north.

Sunday, April 14, 1912

Overnight temperatures had remained fairly mild, around 55° F–60° F, with generally cloudy skies and brisk southwest winds of 15–20 knots. Sunday morning the *Titanic* crossed the cold front that had been just to its west all night. Some scattered shower activity was associated with the front, but again there is no mention of the *Titanic* encountering showers.

A marked change in the weather was behind the front, with brisk northwest winds of 20 knots. Temperatures started to drop from the relatively mild upper 50s to about 50° F by noon. Temperatures would fall steadily through the afternoon and into the night. By 7:30 P.M., the temperature was down to 39° F, setting the stage for the coldest night of the voyage.

The Night of the Sinking, April 14–15, 1912

By 10:30 P.M. Sunday evening, the air temperature had dropped below freezing (31° F). Skies were clearing, and the brisk north wind was diminishing. A large Arctic high (central pressure 1033 mb) was situated over Sable Island, off Nova Scotia, slowly moving southeastward toward the *Titanic*.

At 11:40 P.M. April 14, 1912, the *Titanic* struck an iceberg at 41° 43' N, 49° 56' W. Skies were clear at the time, with no moon, and the sea was "like glass." A smooth sea can only occur with a calm wind, which would be expected at night under an Arctic high.

Some say that the lack of wind was a factor in the *Titanic*'s sinking. They maintain that even a slight wind would have caused some ripples at the base of the iceberg, making it easier for the lookout to spot, and perhaps providing enough warning to have avoided the collision. We'll never know for sure whether a breeze would have helped the lookout sound a warning. But what did happen became the most often retold and best-known maritime disaster story in history, even more than 80 years later.

Under a cold, clear sky ablaze with thousands of stars, the *Titanic* sank into the frigid north Atlantic at 2:20 A.M. Monday morning, April 15, 1912. Of the 2,200 passengers and crew, only 700 survived. Cast into the cold night aboard the all-too-few lifeboats, survivors would have to wait two hours in freezing temperatures before rescuers reached them. By the time the *Carpathia* arrived around 4:30 A.M., the survivors were numb with shock and cold.

One survivor tells of a breeze that came up out of the southeast around dawn to add to the morning's chill. This would be consistent with the Arctic high moving east of the site. In fact, photographs of the rescue that morning clearly show small wavelets on an ocean surface that had been mirror-like just a few hours before. Two days later, the shaken survivors finally landed in New York—a journey begun in hope had ended in mourning.

—By Robert Paola. Reprinted with permission from *Weatherwise*.

CHAPTER 9: MARINE WEATHER

Marine Weather Products

Marine Barometer
Robert E. White Instruments, Inc.
34 Commercial Wharf
Boston, MA 02110
(800) 992–3045

Elegantly executed in rich mahogany with silvered dial and solid-brass fittings, this mercury barometer is copied from a fine eighteenth-century original. Designed for use at sea level in a ship's cabin, it's mounted on a brass gimbal to keep it vertical in heavy seas. It accurately measures barometric pressure while providing a handsome addition to your home. It's truly a collector's item, superbly crafted by one of the most renowned makers of barometers in the United Kingdom. An ideal gift for the discriminating yachtsman, it is priced at $1,250.

Fishing Barometer
Wind and Weather
The Albion Street Water Tower
P.O. Box 2320
Mendocino, CA 95460
(800) 922–9463

Track the barometric pressure and trends and predict the probability of good or poor fishing with this durable travel barometer. It comes in a black plastic case, only 3¼ inches in diameter, with a metal carrying ring that's perfect for securing the instrument with a cord. An altitude dial on the back permits elevation adjustment. This barometer is priced at $37.50 plus $1.50 shipping and handling.

Recording Marine Barograph
Robert E. White Instruments, Inc.
34 Commercial Wharf
Boston, MA 02110
(800) 992–3045

You can get hands-off recording for an entire year with the unique Wempe marine barograph. It includes a 365-day strip chart and a one-year pen, plus four "C" cell batteries. The barograph can sit on a shelf or be wall mounted. You get your choice of a teak or mahogany shockproof case that makes it ideal for use on a boat. It is priced at $995.

Marine Publications

Lots of sailing and yachting magazines are available that have regular articles on weather and are a good source for more marine weather products. While these articles naturally are geared toward sailors, anyone who wants to understand the weather better can learn from them. Sailors have been keenly interested in the latest weather advances for thousands of years, and the advances are coming more rapidly now than anyone would have imagined even twenty years ago. Each of these magazines has hundreds of advertisements that will help you keep up to date with the latest weather products and services and learn how to find them and use them.

Ocean Navigator
Navigator Publishing Corporation
18 Danforth Street
Portland, ME 04101
(207) 772–2466

Published eight times each year, this magazine is for serious oceangoing sailors. Articles cover long-range navigation and weather forecasting. Single issues are $3.50; subscriptions are $20 per year.

Sail
Sail Magazine
P.O. Box 56397
Boulder, CO 80321–6397
(800) 745–7245

Published monthly by Reed Publishers, this magazine is geared toward wind-powered vessels. You can expect to find a lot of wind instruments and articles about predicting winds and weather. Single issues are $2.95; the subscription price is $23.94 for twelve issues.

Power and Motoryacht
Cahners Publishing
475 Park Avenue South
New York, NY 10016
(303) 447–9330

Published monthly to appeal to the opposite crowd as *Sail*, *PMY* is written for powerboaters. The weather focus is extensive but different, having little regard for the subtleties of the wind. Single issues are $3.95. The subscription price is $19.95 for twelve issues.

Yachting
Yachting
2 Park Avenue
New York, NY 10016
(212) 779–5300
fax: (212) 725–1035

In circulation since 1907, you'll find everything about classic sailing boats in this premier sailing magazine. *Yachting* readers demand the hottest, most sophisticated equipment available. It will help you keep up with the state of the art. It's published monthly at a single-issue price of $3.00. For subscription information call (800) 999–0869 or send e-mail via CompuServe to 71230,1466.

CHAPTER 9: MARINE WEATHER

Electronic Marine Information

SEAFAX
Stevens Engineering Associates, Inc.
7030 22th Street, SW
Mountlake Terrace, WA 98043
(206) 771–2182

If your vessel is equipped with a shortwave radio, you can get weather information via facsimile anytime, anywhere in the world. SEAFAX processes shortwave digital weather signals and transforms them into fax signals. It then prints the results on any PC printer with an IBM-compatible serial port. The digital processor is $995. A serial printer port is a $250 option.

Marine Fax
OFS WeatherFax
6404 Lakerest Court
Raleigh, NC 27612
(919) 847–4545

OFS WeatherFax II is an advanced hardware and software package that enables you to capture and display high-quality weather-satellite imagery and HF Marine fax on an IBM-compatible personal computer. When attached to the audio output of your receiver, it captures pictures directly from orbiting satellites, providing immediate NOAA sea-surface temperatures and high-resolution cloud-cover images.

WeatherFax uses standard two-line satellite tracking data; a world-map database; and a true, elliptical orbital model to draw the geographic-map overlay. It also displays latitude and longitude grids, geopolitical boundaries, states, islands, and lakes, plus a "you are here" marker that greatly simplifies its use when navigating at sea. Another critical navigation aid is its *Distance Measuring Tool*, which displays the range (in both statute and nautical miles) and the compass bearing between any two waypoints.

Alden MarineFax TR-IV
Alden Electronics
Washington Street
Westborough, MA 01581
(617) 366–8851

This weather tool offers three modes of output, making it a complete marine weather information center. In addition to marine weather fax capability, it includes Navtex for emergency weather and navi-

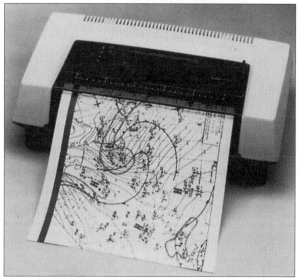

gational bulletins as well as nonemergency information and a radioteleprinter (RTTY) for a broad selection of both marine- and land-based news and weather services.

It includes a programmable memory that will store weather information taken automatically from the sites and at the times you specify, for unattended operation. It also will collect Navtex messages simultaneously with faxes and RTTY data. Emergency Navtex messages are received and printed immediately, overriding all other programmed reception.

Oceanographic Tables
Applied Environmetrics
P.O. Box 241
Roslyn, WA 98941–0241
(509) 649–2940

These tables, on IBM-PC disks, combine the traditional form of scientific tables, used by most scientists, with the flexibility advantages that PC software offers. The tables include ocean dynamics, tidal tables, waves, dissolved gases, sedimentation, physical and transport properties of sea water, geodesy, sunrise–sunset times, the Beaufort scale, blackbody radiation, and statistical tables. An attractive hardbound manual is included.

Applied Environmetrics also offers meteorological tables that provide the reference atmosphere, meteorological temperatures, psychrometry, hygrometry, vertical wind profile, radiation incident on slopes, Gaussian plumes, physical and transport properties of air, deposition, geodesy, sunrise–sunset times, and more. The oceanographic tables are $124.95, but they sell both sets for $199.

OceanRoutes
Weather Network Division
680 West Maude Avenue
Sunnyvale, CA 94086
(408) 245–3600

Here's an online weather service designed for professional use. Shipping companies and professional meteorologists are the prime clients. Usage requires a $15 setup fee and a commitment for minimum monthly connect charges, which cost $1.50 per minute for a 2400 baud modem. The service is accessed through a phone-based network, established by McDonnell-Douglas, called TYMNET. The OceanRoute service provides information from the National Weather Service, the Federal Aviation Administration, and the National Oceanic and Atmospheric Administration.

National Marine Fisheries Service
Northeast Region
14 Elm Street
Gloucester, MA 01930

Southeast Region
9450 Koger Boulevard
St. Petersburg, FL 33702

Northwest Region
7600 Sand Point Way
Seattle, WA 98115

Southwest Region
300 South Ferry Street
Terminal Island, CA 90731

Alaska Region
P.O. Box 1668
Juneau, AK 99802

This extension of the National Weather Service is the central source in the United States for weather information that affects the fishing industry: international fishes monitoring, endangered-species studies and tracking, enforcement, resource management, regulations and investigations, and information storage and management.

Chapter 10
Weather for Pilots

In chapter 9, "Marine Weather," we learned to appreciate how much respect mariners must have for the sea to survive the weather it can whip up. As crucial as the weather is to sailors, there is one group for whom weather is even more critical: aircraft pilots.

If ship captains err in calculating their fuel needs, they will face an embarrassing tugboat tow into the harbor. But pilots who run out of fuel cannot float adrift awaiting help—they and the ground soon will meet, one way or another.

Sailors may be caught by a storm they expected to beat and have to pull quickly into a harbor or drop anchor and hope to weather its passing. But pilots who run into a storm they expected to beat could be tossed against the ground, and they certainly cannot drop an anchor and wait!

There's an old adage about flying that often is seen hanging in flight-school offices at small airports. It usually is the caption to an old sepia-toned photograph of a biplane hanging precariously in a treetop after a crash-landing:

> Flying in itself is not inherently dangerous,
> but to an even greater extent than the sea,
> it is unforgiving of errors.

Pilots must be ever vigilant of the weather. Preflight planning often begins several days before a flight, as pilots check general weather patterns to become aware of what patterns are affecting the country overall. The day before a flight, they may check local weather forecasts for both the intended departure and arrival airports. Just before retiring at night, another check can help pilots get mentally prepared for the next day. The morning of the flight, pilots usually get one more check. Finally, just before departure a final check of official current weather and forecasts for all intended airports is required by federal law before departure.

Until recently there was a vast "class" difference in the weather information available to professional military and airline pilots versus the information available to pilots of small private and commercial aircraft. Let's look at the huge disparity that existed only a few years ago.

Pilots flying the "big iron" of fast and heavy jets got their weather briefings at a full-service weather counter. A professional meteorologist prepared a complete weather package that covered all areas and airports of the pilots' planned flights. Then, when the pilots reviewed this extensive package, they could supplement its information with a wall-size display of radar summaries and weather maps from all over the country. Any questions they may have had were directed to the meteorologist leaning over the counter, helping them interpret this vast wealth of weather information. Thus, the most experienced, weather-savvy pilots flying the planes best equipped to handle any weather got complete, professional weather information.

On the other side of the airport, pilots of small private craft used to be limited to weather briefings by phone. There was no preprepared package covering every aspect of the flight in detail. No weather charts. No radar summaries. No professional meteorologist to interpret. So the pilots who needed it the most—the least-experienced pilots, who may have just been learning about weather and who flew the aircraft most susceptible to poor weather conditions—were the ones who got the poorest-quality weather service.

Fortunately, modern computer technology has evened the playing field. Today every pilot can have a complete, professional weather briefing—every possible chart, summary, and radar picture—directly zapped electronically to any fax machine or personal computer. We still haven't mastered the technology to "zap" the meteorologist electronically, but any Star Trek fan knows that one day pilots may be able to "beam in" a personal forecaster. Then again, that may never happen, because starship pilots don't care much about the weather on Earth.

While weather forecasting has never been better, some weather remains too elusive to forecast accurately. Sadly, one weather phenomenon that has not surrendered its danger to modern forecasting methods is one of the most hazardous: microbursts. Microbursts are a weather event that no one, not even "big iron" pilots, can fly through safely. Still, even these dangerous ghosts of the sky may one day be tamed. Let's take a look at what they are and how we may one day conquer them.

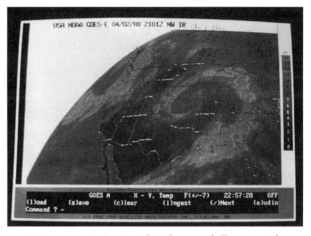

Now even private pilots have a full range of professional weather services.

Microbursts and Airplanes Don't Mix

Twenty years ago, the word "wind shear" was just a fancy term tossed around in meteorological journals and applied engineering studies. But today, it's not the thing you'd want to mention to the average airline passenger. With crowded airspace, busy airports and more airplanes flying in the last twenty years than ever, microbursts have become an increasingly significant menace to aviation, and we are just now finding out how dangerous they truly are. Every now and then, they lash out and claim victims, reminding us that the skies are not ours alone.

Setting the Stage

Some microbursts are caused by cumulus clouds, but others are created by full-fledged thunderstorms. We will now see how this happens.

When a cumulonimbus tower builds skyward, it's not merely "growing." It's being propelled from within by a convective updraft, a strong pocket of warm, moisture-laden air that usually has been heated from below by warm, sunlit ground. As it rises, it condenses and forms cloud droplets. The condensation process also releases small amounts of "latent heat," adding to the buoyancy of the updraft.

As the updraft passes the tropopause and enters the stratosphere, a layer of warmer air 5 to 10 miles above the ground, it begins fighting an uphill battle. Since the surrounding atmosphere has grown warmer instead of colder, the updraft loses its buoyancy, and is inevitably forced to stop in its tracks. Since more air is coming upward from below, the updraft has no choice but to spread horizontally. Looking at the cumulonimbus tower, we see its upper portions begin spreading into the familiar anvil shape.

Birth of the Downdraft

The cloud droplets within the spreading anvil are blown downwind by the existing upper-level winds, which in turn mix much drier air into the anvil. This allows the cloud droplets to start evaporating, changing from droplets to vapor. Since it is the reverse of condensation, latent heat is "taken back," chilling the air, adding to its density and causing it to sink. The downdraft is born! Cloud droplets skirting the edge of the updraft cross over into the downdraft, fueling more evaporation and causing faster downward motion. Larger droplets, being more difficult to evaporate, instead coalesce and combine with each other, forming larger rain droplets. They intensify the downdraft by literally dragging the air down with them.

Once the downdraft reaches the surface, it spreads out horizontally. As the winds pick up, temperatures drop and rain begins, people on the ground head for cover. But if the downdraft is intense, it can send people running for their lives.

The Downburst

Sometimes the relative density of the downdraft can be so strong that it acquires tremendous amounts of momentum and plummets earthward at phenomenal speeds. When the downdraft reaches the ground and spreads out horizontally, winds can kick up to 100 mph or greater, especially on the side ahead of the storm's motion (downdraft winds are always added to storm motion). An intense downdraft that creates damage is known as a "downburst." Fortunately, as the downburst spreads out laterally, its energy thins out and it weakens, confining damage to several square miles.

Damage from downbursts is very similar to that of a tornado. In certain cases, it can be worse. The main difference is that tornadoes involve intense rotational convergence, while the downburst is a divergent phenomenon. Therefore, after a nighttime event, damage survey teams easily can assess whether a tornado or "straight-line winds" occurred based on the damage pattern. It is quite common for residents of a devastated area to insist that they had been in a tornado, but the downburst is no wimp—it can produce winds of 150 mph and roar just as much as the strongest twister.

There are two recognized types of downbursts, classified according to the size of the damage area. The first type is the "macroburst," a downburst that has a damage footprint in excess of 4 kilometers (2.5 miles). Naturally it is the most common type.

Sometimes the downburst concentrates itself into a small, intense pocket as it forms, creating a damage footprint of 4 km or less. Such downbursts are known as "microbursts." Physically, a microburst is no different from its cousin, but since intense wind shear (differences in wind speed and direction) is concentrated in a smaller area, the microburst is much more dangerous to aircraft.

A Menace to Aviation

To the average airline passenger, it seems that takeoff and landing speeds are wild and uncontrolled. But in reality, the pilot is dealing with speed as precisely as if he were driving a car down a highway. As a jet makes its final approach to a runway, the aircraft is flown at the safest, slowest speed possible. Any faster, the plane would land roughly and careen past the end of the runway; any slower, the plane would be too sluggish to control and could even stall.

Unfortunately, the critical speed factor in an airplane is not ground speed; it is the speed of air moving past the plane. That's why it's called airspeed. And when the downdraft falls in front of an

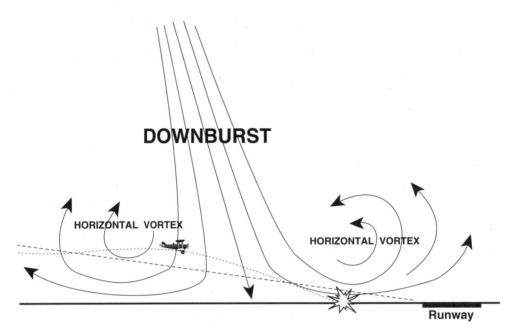

Cutaway Diagram of a Downburst. The aircraft attempts to follow the dashed line, but wind shear plays havoc with control of the plane. Following the dotted line, it encounters a strong tailwind and descends to its demise.

airplane, it creates a wind shear pattern that can interfere with the precise, finely tuned aircraft maneuvers and spell trouble. These areas of dangerous wind shear are strongest in the lowest 300 feet of the atmosphere, where the plane finds itself closest to the ground and is most vulnerable.

When a plane approaching a runway flies into a spreading downdraft, a strong headwind hits the airplane, increasing airspeed and causing excess lift on the wings. The plane rises. To correct this, the pilot decreases thrust to reduce airspeed and counter the lift. As the plane flies directly under the downdraft and moves into the other side of the downdraft, the headwind disappears and a strong tailwind picks up. Airspeed suddenly drops, wing lift decreases and the plane begins descending rapidly. Sometimes the plane strikes the ground before the pilot can recover from the sink induced by the tailwind.

That is exactly what happened in August, 1988, at Dallas-Fort Worth International Airport. A Delta Airlines L-1011 jet was on final approach, descending through a downburst that was positioned between it and the runway. The pilot first reacted to the downburst from in front of the plane, throttling back as he encountered the headwind. But a few seconds later, a tailwind was blowing and he could not quickly get control of the aircraft. The plane dragged across a highway and crashed into a field half a mile short of the runway, killing nearly 150 people.

A similar incident happened July, 1982, when a Pan Am Boeing 727 crashed shortly after take-off at New Orleans, Louisiana. It had flown directly into a downburst off the edge of the runway and descended uncontrollably into a neighborhood, killing 152 people.

Doing Something About It

Fortunately, the aviation community knows much more about downbursts than they did ten years ago, thanks to weather research and education. Weather forecasters have found features on the sounding chart, such as dry midlevel air, that can help sustain strong downburst winds.

Many large airports already have invested in wind-shear detection networks, that alert air traffic controllers to the presence of downbursts and microbursts. This helps the controllers, who now

CHAPTER 10: WEATHER FOR PILOTS

are trained to recognize the visual features of a downburst. Researchers also are trying to find out how Doppler radar might have a place in the airport.

Newer tactics and training for pilots also have made flying safer. Flights through downbursts now are a part of simulator training. For example, most airline simulators have been programmed to recreate the conditions that caused both the Delta and Pan Am crashes. With the advances in training that have occurred since those two crashes, most pilots can safely fly through the once-lethal downburst conditions.

In the air, pilots are strongly encouraged to report wind shear promptly, which helps air traffic controllers and gives other pilots a heads-up. And of course, the pilot's training has encouraged him to use good judgment and common sense, especially when dealing with powerful storms.

Until we find a way to destroy the downburst, there's no better plan.

—By Tim Vasquez of the AAWO.
Reprinted with permission from the *American Weather Observer*.

Here's a personal weather account from Air Force pilot Gary Garber that he experienced while flying the C-5A Galaxy transport jet for the U.S. Air Force at Dover Air Force Base, Delaware.

I experienced the classic wind-shear encounter going into Dover AFB one dark and stormy night with the weather reported at 200 and one-half. Unfortunately, on approach, the field went below minimums. Our official, filed alternate was Wright Patterson AFB in Dayton, Ohio. It was reported clear and a million. But that was nearly an hour away and McGuire AFB in New Jersey—only 20 minutes away—was reporting 400 and a mile and calm winds.

The approach was an arc to final. I overshot the turn to the left of course. This was certainly unexpected with calm surface winds, but it made me aware that we had a strong direct crosswind from the right. Strong winds in the approach pattern and calm winds on the ground should alert any pilot to the potential for wind shear, and we were now ready for anything.

Passing 1,000 feet above the ground, my muscles flexed for the anticipated go-around since the wing-tip winds hadn't decreased. Everyone's antenna was up at this point. Then the bottom dropped out. We lost 500-600 feet nearly instantaneously and just as instantaneously, I was bending the throttles forward against the stops going around.

Three hundred tons of jet and four huge turbofans were barely a match for the wind shear. The Galaxy started shaking so badly that no one could read any instruments, but we remained airborne. My copilot glimpsed the field as we bottomed out. After things settled down, we advised air traffic control to warn another jet on approach behind us. At this point, my only option on the east coast, based on weather and fuel, was Pease AFB in New Hampshire. When we arrived, it was about 600 and 2 in a driving rain storm, but I practically kissed the ground after I landed. I pulled a beer from my flight bag and climbed on the crew van—the beer was history before I finished filling in the log.

Lightning Detector

Small aircraft pilots have long yearned for airborne radar systems to detect thunderstorms in-flight. But radar is too expensive for the budgets of most private pilots and, besides the cost, few small aircraft are well suited for radar. Many private pilots today, however, can get fairly reliable thunderstorm detection by using airborne lightning detectors. These instruments are inexpensive, lightweight and have low power requirements, making them available to large numbers of private pilots.

Now, even nonpilots can enjoy this technology with a ground-based lightning detector from Airborne Research Associates. Their models F-10 and M-10 are designed to warning of lightning in the area long before you see it, thereby giving you plenty of time to take precautions to prevent injury or property damage.

More people are killed or injured in the United States each year by lightning than by tornadoes, floods, and hurricanes combined. Most are not struck directly, but rather are injured or killed by being within a few hundred feet of a strike.

Airborne Research Associates also makes a ground unit for nonpilots

Lightning induces deadly electrical currents in nearby grounded objects, and people can be ground objects, as can construction cranes, livestock, and boats.

The primary difference is that we have adequate warning systems for those latter killers, but few people ever get any warning about impending lightning strikes. But deaths and injuries can easily be prevented through early detection of lightning. Lightning detection capability enable people to continue play or work even when clouds may look ominous. And, sometimes, detection will warn you to take precautions even thought the clouds appear benign. The Airborne Research detectors offer adequate warning times to clear a golf course, stop fuel transfers, shut down cranes, get off ladders and bring children in from playing fields.

For more information, contact Airborne Research Associates at (717) 890–8381.

Riding the Jet Stream: Friend or Foe?

In Chapter 3, "Violent Weather," in the account of "The Storm of the Century," you learned how the jet stream can cause a developing low-pressure system to run amok and create a monster storm. If you ride on commercial jet aircraft, you may have received the pleasant treat of being interrupted by the pilot over the loudspeaker to inform you that the time of the trip will be shortened considerably because of a strong tailwind from the jet stream.

A *jet stream* is a strong wind—usually blowing from the west or the adjacent direction of northwest or southwest—compressed within a thin stream in the atmosphere that wanders around the globe. It can range in width from 25 to 100 miles and can be up to 2 miles in depth. The jet stream is usually broken and splits at several points, but it can remain as a continuous band. The jet stream moves to different elevations all over the world, and its path moves much like a wave.

Although there are jet streams in both the Northern and Southern hemispheres, the most prevalent jet stream, and the one we know the most about, is in the Northern Hemisphere. It is commonly referred to by the term "jet stream," sometimes shortened by pilots simply to "the jet." It is situated in the high tropopause, next to the polar-front zone, where the extreme horizontal temperature contrast exists. Wherever the jet stream is there usually is a break in the tropopause; in other words it occurs where the tropopause has its greatest slope.

The jet stream is stronger in winter than in summer, and the winds forming the band must reach a

CHAPTER 10: WEATHER FOR PILOTS

minimum of 50 knots (58 mph) to be called a jet stream. Most of the time the winds range from 100 to 150 knots (115 to 173 mph), but the high-velocity winds in the jet stream often exceed speeds of 250 knots (288 mph).

This special air circulation, or jet stream, is very important to understanding weather because it is closely associated with migratory low-pressure systems and the polar front. The jet stream aids meteorologists in forecasting weather relative to the development and movement of fronts and low-pressure systems.

The jet stream has a tremendous impact on both weather and flight planning. Aviators must track the jet stream to know if it is going to be a friend and give them an extra "push" or a foe and require an extra stop for fuel.

Still, advanced flight planning is only one source that high-flying jet pilots use for jet stream information. The jet stream can meander far and wide, come and go, and even reverse course, flowing from east to west. Jet-stream forecasts are but rough estimates; the best source of accurate jet-stream data is pilot reports from other aircraft.

Pilots are required by aviation law to report all weather conditions that deviate significantly from the forecast. The jet stream often does, so pilots routinely radio "winds aloft" reports to air-traffic controllers. A pilot who obtains regular weather updates and asks for winds-aloft pilot reports will rarely be caught short of fuel because of a jet stream that is much stronger than forecast.

In flight planning, however, you've got to keep "first-things-first"; the number-one thing in flying is a top-notch, initial weather briefing. Here's an introduction to an excellent source for the kind of up-to-the-minute information—including the jet stream—that is needed to plan any aircraft flight, followed by a listing of a host of other sources for aviation-weather reports, forecasts, and instruction:

DUATS

GTE DUATS
15000 Conference Center Drive
Chantilly, VA 22021-3808
(800) 345–3828
modem: (800) 767–9989
http://www.gtefsd.com/avaiation/GTEaviation.html

DUATS is a complete flight- and weather-briefing service available to all PC users with a Hayes-compatible modem or an Internet connection. While DUATS specializes in aviation weather, it has a "plain language" weather mode that even nonpilots will appreciate. Here's the main menu so you can see the services that are available:

DUATS Main Menu

Weather Briefing	1
Flight Planner	2
Encode	3
Decode	4
Modify Screen Width/Length	5
Golden Eagle Services™	6
Service Information	7
Extended Decode	8
FAA/NWS Contractions	9

Each pilot is limited to twenty minutes per session, but there is no charge for basic services. The Golden Eagle Services requires you to enter a credit card number, and you will be charged by the minute; but Golden Eagle time does not count against your free twenty minutes.

Once you're logged onto the system, select choice number 1 from the Main Menu, and then you can navigate it efficiently using the "Quick Path" commands, which take you straight to your desired information. Quick Path commands are three-letter strings: flight-plan services begin with the letter *F*, weather services begin with *W*, and all plain-language weather services begin with *P*. This is a listing of the DUATS Quick Path weather commands:

Aviation Weather Briefings:
WLO—Low Alt Weather Brief
WIN—Inter Alt Weather Brief

WHI—High Alt Weather Brief
WRT—Route Brief with Selected Report Types
WLN—Local Brief Standard Briefing
WLS—Local Brief with Selected Report Types
WSE—Selected Location Weather
WSC—State/Collective Weather
WDR—Defined Radius Weather

Plain Language Weather:
PLO—Plain Low Alt Weather Brief
PIN—Plain Inter Alt Weather Brief
PHI—Plain High Alt Weather Brief
PRT—Plain Route Brief with Selected Report Types
PLN—Plain Local Brief Standard Briefing
PLS—Plain Local Brief with Selected Types
PSE—Plain Selected Location Weather
PSC—Plain State/Collective Weather
PDR—Plain Defined Radius Weather

The best way to learn about DUATS is to call and try it yourself. You can access it directly on a PC by using any standard communications software and calling (800) 767–9989 with your modem. Use 8 data bits, 1 stop bit, no parity, and any baud rate up to 9,600. Perhaps the most efficient way to access DUATS, though, is with one of the software packages reviewed in the product listings at the end of this chapter. DUATS is standard for all aviation software applications for PCs.

Sea-Level Standard Day

In aviation you'll see a lot of references to a term for a "model atmosphere" called the *sea-level standard day*. Aircraft-performance specifications are built around how an aircraft will perform under the conditions of this model atmosphere. We've listed the specifications for that famous day in August 1959 when the model atmosphere became the standard against which all other days since have been measured:

Composition	Amount
Nitrogen	78.09%
Oxygen	20.95%
Argon	0.93%
Carbon dioxide, etc.	0.03%

Altitude (ft)	Temperature (°F)	Pressure (lb/ft^2)	Gravity (ft/sec^2)	Speed (ft/sec)
0	59.0	2116.20	32.174	1116.40
5,000	59.0	1760.90	32.159	1097.10
10,000	59.0	1455.60	32.143	1077.40
20,000	59.0	973.27	32.112	1036.90
30,000	59.0	629.66	32.082	994.85
50,000	59.0	243.61	32.020	968.08
75,000	59.0	73.78	31.944	968.08
100,000	59.0	23.08	31.868	1003.20
200,000	59.0	0.47	31.566	1038.70

Aviation Weather Products

Weather Videos, Textbooks, and Courses

Sporty's Pilot Shop
Clermont County Airport
Batavia, OH 45103
(800) 543-8633

Most items in Sporty's catalog are strictly for pilots. Still, it has some terrific products that any weather enthusiast will enjoy. Request a catalog and check out its weather video. Since it emphasizes flying weather, you'll get in-depth coverage of thunderstorms, clouds, fog, and ice. It also has textbooks and videos for complete pilot courses and general weather books targeted to help you understand how to predict weather. Here's a sample of Sporty's weather-book titles:

- *Weather Ways*—Covers basic weather theory such as clouds, pressure, wind, icing, thunderstorms, and mountain waves. From the Canadian Department of Transportation.
- *Aviation Weather for Pilots*—The official U.S. government guide to weather, a must for all student pilots.
- *Aviation Weather Services*—A supplement to the above book that can tell you all about how and where to obtain weather information.
- *The Pilot's Guide to Weather Reports, Forecasts and Flight Planning*—A guide to interpreting surface observations, surface analysis, and weather-depiction charts as well as local surface forecasts. Lists U.S. locations of where to obtain surface reports or forecasts. By Lankford.
- *A Field Guide to the Atmosphere*—A virtual picture book, filled with color photographs, that thoroughly explains the atmosphere. From the Peterson Field Guide series by Schaefer and Day.
- *Microbursts: A Handbook for Visual Identification*—An excellent pilot resource for learning to avoid microburst hazards. From the U.S. Department of Commerce.

Aviation Magazines

Many aviation magazines are on the market today. All the following ones have regular articles on weather. While these articles are geared toward pilots, anyone who wants to understand the weather better could learn a lot from them. Aviation always has been on the leading edge of weather advances. So you'll find a host of advertisements in all of these publications that will help you keep up to date on the latest weather products and services and learn how to get them.

Flying

Hachette Magazines, Inc.
1633 Broadway
New York, NY 10019
(203) 622–2700
fax: (203) 622–2725

Published monthly, featuring articles for all pilots, including private, commercial, and airline transport flying piston-engine aircraft, turboprops, or jets. Single issues are $2.95. The subscription price is $24 for twelve issues in the United States, $31 in Canada, and $32 in all other countries.

Plane and Pilot

Werner Publishing Corporation
Suite 1220
12121 Wilshire Boulevard
Los Angeles, CA 90025–1175
(800) 283–4330

Published monthly and targeted for private and commercial pilots of all sizes of piston-engine aircraft. Single issues are $2.95. The subscription price is $16.95 for twelve issues.

Private Pilot

Fancy Publications, Inc.
3 Burroughs
Irvine, CA 92718
(303) 786–7306

Published monthly and written for private pilots flying small private aircraft. Single issues are $2.95. The subscription price is $23.97 for twelve issues.

Electronic Aviation Weather Products

Weather Fax

Weather Fax, Inc.
52 Domino Drive
Concord, MA 01742
(800) 359–4242

If you have access to a fax machine, you can use this service to receive fast, clear pictures of current weather conditions. Weather Fax can instantly send you sharp, high-resolution weather charts and data that give you easy-to-understand, up-to-the-minute information it has compiled from official National Weather Service data. It's available seven days a week, twenty-four hours a day.

AVFAX

American Flight Service Systems, Inc.
4640 East Elwood
Phoenix, AZ 85040
(800) 432–3265

This service maintains that it has the only "pilot-defined" aviation-route briefings via fax. The additional services it provides are of interest only to pilots; so if you don't fly, you'll receive more information than you need for a weather briefing. Order a tailored route briefing and AVFAX will fax you two weather charts plus SAs, FTs, FDs as well as the pilot-specific information (pireps, notams, sigmets, and airmets).

Flitesoft Professional

RMS Technology
124 Berkley Avenue
Molalla, OR 97038
(800) 533–3211

Here's a lean, easy-to-install pilot-briefing program for PC users. Flitesoft provides a fast, easy, and accurate way to do all of your flight planning. It uses Jeppesen NavData and includes graphic and automatic flight planning, weight and balance checks, weather briefings, and functions to automate every important planning task.

With Flitesoft, weather briefings (from

DUATS) are automatic and free. Winds aloft are inserted into your route automatically, and free weather graphics appear directly on your computer-generated flight chart.

Sample, demonstration program is available on the Internet at http://www2.aero.com/catalogues/software/rmsdown.htm.

The feature that will most interest nonpilot users is the internal DUATS communications program, which enables you to download winds and weather. It then integrates them into your flight planning for time and fuel calculations. It is available in both Windows and Mac versions.

FliteStar

MentorPlus Software
22775 Airport Road, NE
Aurora, OR 97002
(503) 678–1431

This program has some sophisticated flight-planning features, like automatic route selection and a special "flight engineer" window that lets you ask "what if" questions to test various altitudes for wind effects on fuel consumption. Its answers are calculated using weather information downloaded from either Jepp-Link or Contel DUATS.

FliteStar has an excellent manual that is supplemented with a thirty-minute training video. It has a nice display, chock-full of information. This display comes at a price, though; the display almost tries too hard and creates a cluttered look, and it requires at least a 386 computer for adequate performance. Versions are available for both IBM and Mac at $295.

George
AzureSoft
1250 Aviation Avenue
San Jose, CA 95110
(800) 282–6675

This program may be the best of the flight-planning programs if your main interest is in downloading weather information. It's built around another AzureSoft product, Complete Briefer, that is the best DUATS translation program available. The module with George works with both the DTC and the Contel DUATS services.

The flight-planning features in George are a bit weak, especially compared to FliteStar. It does not permit "what if" questions that would truly tap the power of having downloaded a complete set of winds-aloft forecasts.

Weather Monitor II
Davis Instruments
3465 Diablo Avenue
Hayward, CA 94545
(800) 678-3669

Personal-computer users will love using this home weather station with its optional PC interface. The Weather Wizard is an inexpensive, sophisticated, fully automated home weather station that includes inside and outside temperatures, wind speed and direction, windchill, time and date, recorded highs and lows, alarms, metric conversions, and a self-emptying rain collector for less than $400.

Talking Weather Stations
ItWorks
P.O. Box 7403
Chico, CA 95927
(916) 893–6510

This weather station is perfect for small airports. It will generate local weather reports by recording weather data using an IBM-compatible PC and playing it back over any telephone. Completely menu-driven via the PC software, you can program a playlist with custom messages, ads, and reports and handle up to sixteen callers at once. It uses a built-in voice library, or you may record your own using anyone's voice.

The station is totally self-sufficient and maintains call statistics so you can check usage. All weather data are available for your reports, including wind speed and direction; gusts; peak gust with time stamp; windchill; relative humidity; dew point; barometric pressure with trend and rate information; rainfall for day, month, and year; rainfall rate; temperature with maximums and minimums by the day, month, and year with date/time stamps; and current time. A complete two-line system costs $1,950.

Wireless Weather Warnings
Intelligent Information Incorporated
1315 Washington Boulevard
Stamford, CT 06902
(800) 633–8273

Here's a way to get a taste of weather's electronic future, yet someday it's going to seem as primitive as an old rural phone system where you turned a crank to ring up Betty Lou and have her connect you to your neighbor down the road. Still, like those old phone systems, it's a great boon to anyone who needs it today. How does it work?

A new electronic weather source called Weather Alert Service (WAS) is now online and available to anyone with a standard, alphanumeric personal pager. For about $100 per month, you can get nationwide pager service (including the required pager) that will beep you with weather alerts that you've programmed for with "alert triggers."

Regional service is about $30 per month; local service is about $10 per month.

WAS could pay for itself in one use if you're sitting at an airport waiting for destination minimums to come up or if you're en route and concerned that your destination could go illegal at any moment. Your beeper will alert you if any critical weather minimums have been reached, after you've called in your specifications via the toll-free phone number. With the national service the pager will alert you anywhere, even in flight (when below 10,000 feet), thus warning you that you may need to consider your alternatives.

WAS goes beyond simply alerting you to programmed weather alerts, delivering full SAs or FTs or special forecasts that can be read in their entirety in the alphanumeric window. It is not as yet intended for getting a full weather briefing, but that most likely will soon change.

Expect to see this technology evolve into a complete system for notebook computers and portable printers that will enable pilots to print out a full, targeted weather briefing at any time and any place with automatic updates if conditions go below acceptable minimums. But even today it can communicate with Apple Newton Message Pads that are equipped with the optional PCMCIA paging card. Another alternative is the Zoomer "personal digital assistant," produced jointly by Tandy, Casio, and GRID, that also will have a slot for an optional paging card.

Chapter 11
How to Become a Weather Forecaster

Observing the local weather and carefully recording meteorological statistics are a serious pastime for thousands of amateur scientists. In the following article longtime weather observer Steven D. Steinke introduces several of America's 25,000 amateur weather watchers and describes the essential components of a home weather station. Their stories and Steve's expert advice can help you take your first steps toward learning how to become a weather forecaster.

Bill Canning is a travel agent from Toronto, Ontario, Canada, who is fascinated by the silent beauty of a surprise snowstorm.

Debi Iacovelli is a housewife from Cape Coral, Florida. She keeps a complete and detailed computerized weather log for every day of the year. Debi sometimes composes lines of poetry to help her recount a particular event.

Clint Martz is a sixteen-year-old high school student from Elmira, Idaho. He watches weather like most people his age watch football games.

And Roy Britt works for the Nabisco cookie bakery in Richmond, Virginia. Roy spends some of his vacation time each spring chasing tornadoes along the endless dusty roads of west Texas and Oklahoma.

Every day these individuals interrupt their daily routines to step outside, expose all their senses to the state of the atmosphere and fall in love with what they see, hear, and feel. They are watchful observers, diligent record keepers, and eager and appreciative students of the atmospheric elements around them.

Why the Intense Interest in Weather?

America's serious weather watchers, perhaps 25,000 or more in number, range in age from as young as 6 or 7, to a full century older. Many become weather watchers after a particular weather event affected them personally: Chicago's big snow of 1967, the Big Thompson Canyon flood of 1972, the great Long Island hurricane of 1938, or, as in my own case, the disastrous Belvidere tornado of 1967. Each event had a major impact in the shaping of a new weather observer.

Today, nearly 11,000 of America's dedicated observers directly serve the National Weather Service as members of the National Oceanic and Atmospheric Administration's Cooperative

Observer Program headquartered in Silver Spring, Maryland. Some provide a variety of information several times each day, while others methodically and meticulously maintain daily and monthly rainfall, snowfall, temperature, or river level information for this country's permanent climatological record.

Assembled and archived at NOAA's National Climate Data Center in Asheville, North Carolina, this significant historical record includes multiple decades of data from thousands of locations under NOAA's jurisdiction. It is a data base produced by people dedicated to a most important element of our historical record—the climatological history.

Thousands of other observers have as their only reward the personal satisfaction found in simply recording the elements of the atmosphere which blow through, fall upon, and run off their own back yards. They seldom seek personal gain or financial reward, but are justly proud of their intimate understanding of the weather outside their doors.

Though many of these backyard observers do not function in any official capacity, they frequently provide valuable and timely information to local National Weather Service offices, television and radio stations, and emergency services during times of threatening weather. Each year some are cited for their contribution to the safety and well-being of the people within their communities.

Weather Organizations

Over the years, several organizations have been instrumental in fostering interest in weather and climate study.

The American Meteorological Society is the patriarch of professional meteorological organizations in the United States. With an emphasis on academics and research, the organization's membership continues to represent the weather and climate professionals in an exceptional manner. Recent educational initiatives by this organization's leadership clearly show its commitment to stronger science education for younger students as well.

The National Weather Association represents practicing meteorologists. The NWA seeks to develop a responsive, motivated, and highly qualified pool of professionals who can be called to interpret, forecast, and assess the weather and climate around the country in a professional and proficient manner.

In spite of the impact that these organizations have had on many of us who watch the weather, until recently there was no organization dedicated to sharing and enhancing interest on an amateur level. This situation was remedied in 1983 by the formation of the American Association of Weather Observers (AAWO), an organization which has provided an opportunity for observers to share their love of the weather with thousands of others. The AAWO has grown from a handful of weather watchers to a membership in 1990 of some 2,000 persons throughout the United States and numerous foreign countries including Argentina, India, The Netherlands, Switzerland and others.

—By Steven D. Steinke. Reprinted with permission from the *American Weather Observer*.

CHAPTER 11: WEATHER FORECASTER

Amateur Weather Organizations

American Association of Weather Observers
(AAWO)
P.O. Box 455
Belvidere, IL 61008
(815) 544-5665

International Weather Watchers
P.O. Box 77442
Washington, DC 20013
Internet: c.geelhart@genie.geis.com
GEnie: c.geelhart

National Weather Association
440 Stamp Road, Room 404
Temple Hills, MD 20748

Interior of Eastern New York AMS
SUNY at Albany
Dept. of Atmospheric Sciences
1400 Washington Avenue
Albany, NY 12222

Chesterfield County Weather Network
P.O. Box 808
Chesterfield, SC 29709

Atlantic Coast Observer Network
92 Village Hill Drive
Dix Hills, NY 11746

Blue Hill Observatory Weather Club
Box 101
East Milton, MA 02186

Long Island Weather Observers
Box 259
Carle Place, NY 11514

Mount Washington Observatory
1 Washington Street
Gorham, NH 03581

Northeast Ohio AMS Chapter
1667 Cedarwood Drive, Apt. 109
Westlake, OH 44145

North Jersey Weather Observers
Box 619
Westwood, NJ 07675

Akin Chapter of the AAWO
Bruce Watson
2514 Brenner Street
Roseville, MN 55113

Long Island Weather Observers
2 Stanley Court
Lake Ronkonkoma, NY 11779

Skywatchers Club of California
c/o Richard Dickert
6607 South Prospect, #102
Redondo Beach, CA 90277

Pacific Northwest Weather Observers
c/o Robert Kleeman
Star Route Box 360
Spirit Lake, ID 83869

California Weather Association
Richard Dickert
511 Faye Lane
Redondo Beach, CA 90277

Virginia Weather Observers Network
Central Virginia AMS
University of Virginia
Charlottesville, VA 22903

THE WEATHER SOURCEBOOK

How to Become a Home Weather Forecaster

Home weather stations range from the simple to fully computerized operations with satellite photo capabilities and direct access to National Weather Service maps and data. In this example, you will learn how to observe the elements of the atmosphere that are most frequently measured and recorded at amateur stations around the country along with a simple explanation of how to do it.

Precipitation

The primary element of the atmosphere measured and recorded by both amateur and professional meteorologists and climatologists is liquid precipitation. Rainfall can be measured using gauges that range from coffee cans to fully electronic gauges capable of melting snow and ice and measuring its liquid content as it falls.

Both rainfall and melted frozen precipitation should be measured to an accuracy of 0.01 inch. Garden-style gauges, although suitable for many applications outside of serious amateur weather observing are usually graduated in tenths of an inch and often lack the accuracy required for serious observations.

Accuracy of measurement is controlled by the diameter of the catch (opening of the gauge) and its location at the recording site. A 4-inch diameter gauge is more accurate than a 1-inch gauge. Eight-inch gauges are the accepted standard within the National Weather Service. They have smaller diameter measuring tubes within the larger cylinder to provide for a reliable measurement to 0.01 inch. A 4-inch gauge costs about $30.00 while an 8-inch standard metal gauge might sell for $220.00 or more.

It is essential to have the gauge secured to a post or the ground so that it does not succumb to the winds that may accompany the storm. There is nothing more frustrating than having the heaviest rain of the year be lost to a tipped-over rain gauge with no way to recover the measurement.

Snow and Temperature

Snow can be captured and measured in a rain gauge, or it can be measured with a ruler on the ground on a white, hard-surfaced substrate known as a snow board. Snow is usually measured to the nearest tenth of an inch. When keeping a continuous record of total liquid precipitation at a given station, it is important to melt snow and determine its liquid depth.

The second weather element frequently measured and recorded is temperature. Utilizing thermometers of varying degrees of accuracy placed in locations with various types of shielding from the sun, many observers maintain daily records of 24-hour maximum and minimum readings over the course of a year.

The simplest home thermometer can provide the basis for interesting record-keeping. But because high and low temperatures for any given day do not always occur at a time when they are conveniently observed, a good investment for the serious observer is a maximum-minimum thermometer that marks the high and low temperature during the time between observations. The least expensive thermometer costs less than $50.00, while those meeting National Weather Service specifications may run as high as $175.00 or more.

Equally important as the type and quality of the thermometer is its placement. Thermometers must be placed in a location protected from direct sunlight and away from large buildings, blacktop areas, or other areas where the surrounding environment might alter the accuracy of the measurement. Ideally, thermometers should be housed in an instrument shelter with good air flow and no direct solar exposure.

Pressure

A third element often measured is air pressure. Using either the common aneroid barometer or a more expensive mercury barometer, the observer records pressure changes as they occur, noting both increases and decreases as various weather systems pass overhead. The direction and rate of change in air pressure at any given location is perhaps the single most important measurement used in forecasting the next twelve hours of weather.

The most common barometers are the aneroid variety. Most have a clock-type face with two measurement needles. One needle provides a manually adjustable reference marker while the other reflects the change in pressure as measured by the internal workings of the instrument.

A newly acquired barometer must be calibrated. The easiest method is to contact a local weather observer or local National Weather Service office to get a current reading. You then follow the instructions supplied with your instrument to set its reading. It is sometimes necessary to lightly tap the glass face of an aneroid barometer with your fingernail to allow the needle to settle on the correct reading.

After the barometer is calibrated, you should rotate the reference needle until it is directly over the current reading.

When you check the pressure at the next observation time, you will be able to note if the pressure has risen (moved to the right) or fallen (moved to the left), and by how much. The relationship to the previous reading gives important information that can be used in forecasting the weather ahead.

Other Elements Measured

Other elements frequently measured and recorded by the amateur observer are humidity, wind speed and direction, cloud cover, fog density, thunderstorm days, heating and cooling degree days, growing season information, solar radiation, rainfall, acidity, seismology, and a host of other parameters.

About the AAWO

The American Association of Weather Observers (AAWO) has as its objectives education, communication, and cooperation among all weather enthusiasts at all levels, regardless of whether they photograph clouds, monitor only a simple indoor-outdoor thermometer, or maintain a fully computerized weather station with floor-to-ceiling instrumentation.

Today there are some 2,000 members of the AAWO, in every state and numerous foreign countries. Through its monthly publication, *The American Weather Observer*, the AAWO helps others share their enthusiasm for weather watching. The publication contains information provided by many AAWO members, including recent weather events, weather stories, free data publication for their own stations, and observer recognition.

The AAWO hosts an annual meeting each fall where members hear presentations by renowned weather experts. "Observer of the Year" awards are an important part of each year's program.

THE WEATHER SOURCEBOOK

> Meetings have been held in numerous locations, but perhaps the most exciting, from a weather buff's standpoint, was the meeting held in 1990 at the base of Mount Washington, New Hampshire, home of the "world's worst weather." An enthusiastic contingent of observers from as far away as California came together for two days of education, information, and outstanding scenery.
>
> Future meetings are being planned for other areas, including on the rim of the Grand Canyon, the National Climatic Data Center in Asheville, North Carolina and in Washington, D.C. The AAWO seeks to encourage, organize, and promote new chapters throughout the country. Some three dozen other organizations are now affiliated.

It's Easy to Become a Weather Observer

Weather observers are among those fortunate people who understand the special relationship between self and environment. Some are as dedicated as Roy Britt, who spends hundreds of hours each year chasing cumulonimbus clouds across the Great Plains in hopes of a 30-second glimpse of a swirling vortex. But you need only observe the weather in your backyard to better appreciate and understand your relationship with the sky.

As Debi Iacovelli has so aptly observed, "I enjoy being able to really see the sky. I have my head in the clouds all the time!"

Once you, too, get the bug, you may want to use your data to help build national weather databanks. If so, you can become a member of the American Weather Observer Supplemental Observation Network. Write to them at AWOSON, 401 Whitney Boulevard, Belvidere, IL 61008–3772 for details.

—By Steven D. Steinke. Reprinted with permission from *Science Probe!*

Old Faithful Station

Old Faithful geyser is located only a half mile east of this weather-station site in Yellowstone National Park, Wyoming. The fenced-in area helps to keep visitors and bison at bay, along with elk which frequent the area year-round. In the 1988 fires nearly two million acres burned. On September 7, 1991, this site almost burned, too, but the park-ranger observers did not miss their observation as the fire came within a few hundred feet of the site and was fanned by hurricane-force winds. In addition to serving as a volunteer weather site, this location also has automated instrumentation linked directly with the Cheyenne (Wyoming) NWS Forecast Office. Although temperatures here sometimes dip below −50° F in winter, the thermal activity along the Firehole River only a short distance away makes the valley a year-round hangout for wildlife and nearly 100,000 visitors.

Tips on Achieving Precision Accuracy

Accuracy and repeatability of results under identical circumstances are important factors in setting up a home weather station. An often overlooked factor in achieving these results is the exposure of your instrumentation. Weather observation instruments must be exposed in such a manner that they measure the element they are designed to measure and are protected from influences which would contaminate their indications.

For example, the typical window thermometer will give a rough approximation of the atmospheric temperature but its indications usually will be contaminated by heat from the house behind the window that filters through the crevices of the window and its frame as well as heat radiated from the building's wall. In addition, such thermometers will at times be affected by the sun's rays as well as by precipitation, which is not necessarily at the same temperature as the atmosphere through which it is falling. Moreover, the thermometer itself can radiate heat from itself if exposed, for example, to the nighttime sky. It is important, therefore, to minimize the effects of extraneous influences on the indication of the thermometer. It must be protected from radiation to or from the instrument, from heat advected from artificial sources and from precipitation.

To begin with, the National Weather Service urges that meteorological thermometers not be exposed closer than 100 feet to any large radiating surface, such as an asphalt parking lot or a large building. Ideally, the thermometer should be exposed about 4½ feet above a grass surface that emits a minimum of infrared radiation. In addition, the thermometer should be shielded from incoming radiation and from the nighttime sky to which it could radiate heat and take on a temperature below the air temperature. Proper shielding also would protect the instrument from precipitation.

The typical means of shielding meteorological instruments is by housing them in an "instrument shelter." The National Weather Service uses a design that has been in use since before the turn of the century. This design is commonly referred to as the "cotton region shelter" and consists of a double-roofed, louvered wood box painted white and door opening to the north (in the northern hemisphere). Its dimensions are approximately 2 feet by 2 feet by 2½ feet high. Ideally, the shelter should be sited over a grassy surface and at least 100 feet from a large radiating surface and no closer to any object than four times the height of that object above the shelter.

Pre-built cotton region—or medium—shelters cost about $450 but can be improvised. You can use several louvered closet doors, heavily painted with a good, outdoor white paint. The doors are available at most lumber and building supply centers.

Even high-tech electronic temperature sensors need protection, customarily sheltered in smaller units known as thermo-shields. These are usually plastic, louvered cylinders within which the temperature probe is placed. However, the typical "cotton region" shelter can house electronic sensors equally well and this shelter has the further advantage that it can house a thermograph or hygrothermograph and all your weather instruments.

Rain gauges should be exposed on a flat surface no closer to any obstruction than twice the height of that object above the gauge. The top of the gauge should be level. For accuracy, check your leveling with a spirit level across the top of the gauge at opposing 90° angles and adjusting the gauge until both levelling positions indicate level. Rain gauges must not be placed near the edge of a roof or near a cliff because of the influence of updrafts. Low shrubbery, situated around the gauge, is an ideal wind-baffle, provided its height does not exceed twice its distance from the gauge.

The distances cited are for ideal conditions. Don't let an inability to achieve the ideals prevent you from setting up your station. Even National Weather Service stations often are unable to achieve them and use compromise measures. Note all limitations in your instrument exposure in your "Station History" log, a document that records the geographical coordinates and ground elevation of the station and other pertinent data, instrument descriptions and locations, as well as changes in location, etc.

Your station elevation can be determined from a U.S. Geographical Survey topographical map for your area, available at a library or for a reasonable cost at Federal Distribution Centers located at designated neighborhood outlets or at the U.S. Geological Survey, Reston, Virginia 22092.

Whatever compromises in instrument exposure may be required, they should not be such as to render readings totally unreliable or unrepresentative. For example, under no circumstances should a rain gauge be placed under over-hanging tree branches or an instrument shelter within a few feet of a south-facing wall or over an asphalt pavement.

Record Keeping

Normally, one daily observation is taken at a climatological station. This observation includes reading and recording maximum and minimum temperatures for the preceding 24 hours, temperature at the time of observation, 24-hour liquid precipitation, snowfall or frozen precipitation depth, total frozen precipitation on the ground and any other meteorological phenomena present such as fog, thunder, etc.

The National Weather Service uses WS Form B-91 for recording daily observations. This form is the model for the form used by the American Association of Weather Observers (AAWO). A year's supply of these forms is available from the AAWO, P.O. Box 455, Belvidere, IL 61008 for $9.00. If you use the AAWO record forms and return a copy to the above mentioned address, you will be contributing to a databank and performing a valuable public service.

At the end of each month, you should make a statistical summary of the month: average maximum temperature, average minimum temperature and average minimum temperature divided by two, total liquid precipitation, total frozen precipitation, etc. The AAWO form has a monthly summary printed at the base of each monthly form. At the end of the year, temperature and precipitation data should be appropriately summarized.

Helpful Literature

National Weather Service Observing Handbook No. 2, Cooperative Station Observations, Silver Spring, MD 20910, 1989, contains a wealth of valuable information on instrumentation and siting as well as on observing techniques. Your local National Weather Service office should have a copy. Qualimetrics, formerly Science Associates, Inc., has a master distributor, Novalynx Corp. (address: P.O. Box 240, Grass Valley, CA 95945, 916–477–5226) that can supply catalogs of the meteorological instruments they offer (professional and amateur level) and price lists. Your local library can supply you with books and literature that contain plans for building and details on using home weather stations.

—By Rev. Robert J. Duane. Reprinted with permission from the *American Weather Observer*.

Internet Weather Services

The Internet has become a rich source for all kinds of weather and weather-forecasting information. Listing the resources here would be a waste of space. Besides, things change too fast in cyberspace to restrict you to what we could include in this book.

CHAPTER 11: WEATHER FORECASTER

> The only problem you'll have in using the Internet for weather information is that you'll have *too much* information! Use any of the Web search engines, such as *www.yahoo.com*, *www.lycos.com*, or *www.excite.com*, and search for your favorite weather topic. It may take you a while to narrow down the listings to the information that you want, so once you find a good resource that answers your specific needs, be sure to add it to your bookmark list!
>
> If you know how to write Web pages with HTML, you can make your own weather page by assembling all of your favorite weather links. Then, save this HTML page on your hard drive and add just one "Weather" bookmark that will bring up this saved page.

The Weather Channel Online

You don't even need to turn on your television to enjoy the benefits of the Weather Channel. In fact, their online service may better suit your needs. Check the weather anywhere in the world before you take a trip by visiting the Travel Weather department to track delays and find out what to pack. And, during ski season, the "Powder Report" link shows you the best place to go on your next ski trip.

Perhaps the most useful service they offer is regular local forecasts. Look for the United States map, jump to your state, and then to your city for a full report. You'll get current conditions, a five-day forecast with graphics, and a link that will display a local radar image. Once, you've found your local weather, use your Web browser's commands to add the page to your Internet Explorer Favorites Folder or your Netscape Navigator Bookmarks list so you can get instant weather updates everyday.

There's much more here than just forecasts, though, so be sure to explore all the links to get a rich variety of weather facts, news, and regularly-updated special features.

For more information on the Weather Channel, be sure to read out in-depth coverage and interviews in the next chapter.

Forecasting Products

WeatherView
Jim Haywood
2157 Forest Hill SE
Grand Rapids, MI 49546
CompuServe: 72247.2072

WeatherView is a comprehensive weather software package geared for the nontechnical weather enthusiast. Hourly data released by the National Weather Service and Canada via the Surface Airways (SA) reports and Digital Radar (SD) reports provide the raw data needed to use WeatherView. Each user is responsible for providing his or her own SA and SD data. Common cost-efficient data sources include CompuServe, DUATS, and Weathermation.

WeatherView is simple to use, colorful, and quick. Display choices are provided, by menu scroll bars and "hot keys" for instant access. Documentation is provided on disk to maintain low overhead costs. Raw SA and SD data are formatted into a column-oriented format and can be imported into other database-management programs for further analysis. WeatherView allows the user to save and review significant hourly weather events. Editing of data within the program is allowed to correct invalid data.

Hourly NWS data are presented in a graphic display format featuring a national-map overview and zoom mode for detailed analysis. Parameters graphically displayed include current weather, sky cover, visibility, pressure isobars, location of highs/lows, temperature, dew point, wind speed, wind direction, radar, relative humidity, wind chill, and heat index. Graphic display parameters are color-coded and can be customized by each user.

WeatherView allows you to search hourly NWS data by location or weather event. You need only to enter a city name or code to receive the latest information in easy-to-read common text. You also may choose to enter a weather event (such as snow or thunderstorm) to receive a breakdown of all stations reporting that event in easy-to-read common text. WeatherView also provides a list of all stations reporting extreme conditions, such as high/low temperatures, high/low pressures, high wind speed, and extreme precipitations from the current data set.

User configuration includes custom colors for graphic-display parameters, inclusion or exclusion of Canadian data in the search area, isobar smoothing passes, zoom-window isobar spacing, time zone, choice of text editor for data editing, and 3-D shadowing option. A utility program allows the user to add or delete stations to the graphic maps.

Hardware requirements include:

- MS-DOS operating system
- 512K minimum available RAM
- EGA minimum video
- Minimum 80286 processor (80386 or better recommended)
- Math coprocessor recommended

Primary access to the program is through the CompuServe AVSIG forum (library #1). A free demo is provided through this forum (WXVIEW.ZIP) that allows full usage with a provided demo data set and limited usage with user-provided data. Registered users receive a "key" file that removes the imposed limitations. By downloading periodic upgraded demo versions, registered users receive free registered upgrades. The author can be contacted for comments and suggestions. WeatherView registration is $25 plus $4.95 shipping and handling. (Michigan residents add 4 percent sales tax.)

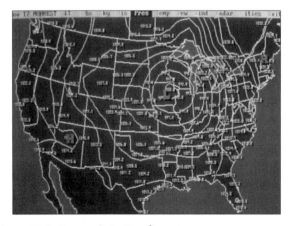

U.S. Barometric Readings

Chapter 11: Weather Forecaster

WS2000 Home Weather Station

RainWise, Inc.
P.O. Box 443
Bar Harbor, ME 04609–0443
(800) 762–5723
fax: (207) 288–3477

The RainWise WS2000 is the latest in private weather-station technology. It includes wireless transmission of sensor information and a solar-panel power system, which enables you to locate it up to 300 feet away from the display unit. Its design just about eliminates all of the problems associated with installation and servicing of a weather station.

It measures, records, and processes a wide range of weather data, displaying all of it on a large LED interface. The WS2000 will give you a professional degree of accuracy, and yet it's priced for a home budget.

- Inside and outside temperatures: both from –99°F to 150° F
- Barometric pressure: 28 to 32 inches of mercury
- Wind direction: in 10° increments
- Wind speed: from 0–150 mph, or in knots
- Windchill: from –127° F to 88°F
- Relative humidity: from 0 to 99 percent
- Rainfall: up to 99.99 inches at 0.01-inch accuracy and with an automatic tipping rain bucket
- Degree-days heating and cooling: stores deviations from the 64° F ideal in memory so you can check your heating or cooling systems

The printing model automatically prints a log of your local weather conditions twenty-four hours a day, giving you a digital record of outside tempera-

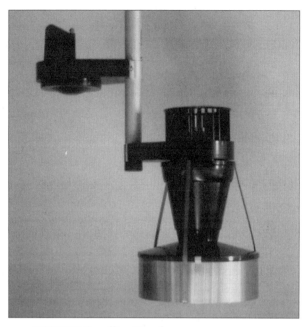

WS2000 Satellite Head

ture, relative humidity, barometric pressure, wind speed, wind direction, and rainfall. The nonprinting model is designed for wall-mounting, but you can order optional feet and keep it on a desktop.

RainWise also makes the WeatherVideo, which displays all the information listed above in color on your television screen. Instead of a traditional LED display box, you can select your own personal "weather channel" and see an instant readout of current weather conditions, even paging between multiple screen options. The unit includes a video processor that connects to your television set and a light pen that you use to select options directly on your screen.

The WeatherCycler

The Weather School
5808 Tudor Lane
Rockville, MD 20852
(301) 230–8985

The WeatherCycler is a portable, all-purpose, pocket weather forecaster. By simply pulling the chart insert, the weather systems are put into their typical movements from west to east. Weather

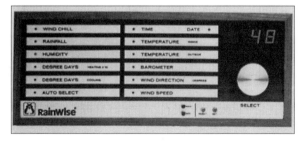

WS2000 Control Panel

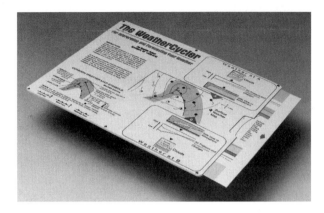

changes are shown for your location as the weather systems pass by. Forecasts can be made and verified instantly by flipping to information displayed in two windows on the back side of the chart. The WeatherCycler is constructed of rugged calculator board with six metal eyelets and is printed in four colors.

Suitable for use by anyone interested in the weather, it's especially useful for pilots who like to maintain an ongoing awareness of the weather. Of course if you're a student pilot, this is an excellent weather-training aid. There's also an instructor's transparency version that can be used on an overhead projector. And for classroom use, The Weather School has a related WeatherCycler Study Activities book that presents weather basics.

Chapter 12
Major Equipment

Fewer than ten years ago, accurate weather observations and forecasting were accomplished only by users of the most sophisticated weather equipment. Today, however, with the advent of personal computers and miniature electronics, nearly anyone can easily obtain a complete home weather station to gather local weather data. Moreover, you can tie into a vast network of up-to-date and accurate weather data from around the world, then feed that data into personal-computer applications that produce forecasts that will rival the experts in accuracy.

Still, most weather aficionados will never personally see or use any of the major types of weather-observation equipment used by large professional or government organizations. In this chapter you'll learn more about the latest, most sophisticated equipment in use today, about how it's used, and about the organizations that use it.

One of the most expensive and most sophisticated weather instruments ever developed is the new NEXRAD Doppler radar systems being built around the country. The systems are already producing some amazing results. For example, in Oklahoma City—the midst of Tornado Alley—one of the first NEXRAD installations has increased the warning time for tornadoes to an average of 19 minutes. Until but a few years ago, tornadoes seemingly struck from nowhere, with little warning. What warnings were given often proved to be false. False warnings build complacency, and people then develop a habit of ignoring all warnings. NEXRAD gives us the technology to issue ample warning time and reduce false warnings. The combination of these two already has saved lives. Here's an update on our new NEXRAD system.

NEXRAD

Weathermen are making bright forecasts for a new radar that can peer into the very heart of violent storms. The Doppler radars now appearing around the United States are expected to save money and lives by providing earlier warnings of severe weather events such as flash floods, tornadoes, and thunderstorms.

"The radar allows us to see what's going on in the atmosphere, even before a storm forms," says Joe Friday, director of the National Weather Service in Silver Spring, Md. "And after a storm gets moving, it can map the distribution and intensity of precipitation inside it."

Also known as NEXRAD, an acronym for next generation weather radar, the Doppler can detect and track both clouds and cloud-free fronts. Severe storms frequently develop along such fronts, and being able to track them can provide hours of advance warning.

"The military pioneered Doppler radar technology in the 1950s by using it to help guide ground-to-air missiles," says Robert L. Dwight, curator of the Historical Electronics Museum near Baltimore.

Civilian radars appeared in the 1960s as navigation aids on commercial aircraft; since then, scientists have put them to many uses, including the mapping of Venus and the study of ocean waves. Policemen use them to catch speeders.

The Doppler effect makes train whistles or car horns change pitch as they approach and travel away from a listener.

A Doppler radar sends out a pulse using a very stable frequency, and measures the frequency of returning signals. The frequency shift is precisely measured by the radar to determine the speed of rain or snow particles, cloud droplets or dust moving toward or away from the radar.

The first three Dopplers are operating in Sterling, just outside Washington, D.C.; Norman, Okla.; and Melbourne, Fla., near the Kennedy Space Center. Others will soon be functioning in St. Louis, Missouri, the Galveston/Houston area of Texas and Dodge City, Kansas.

By 1996 a nationwide network of 175 NEXRADs, costing about $437.5 million, should be scanning the skies from national weather stations, defense installations and Federal Aviation Administration facilities. Each of them will see weather 125 miles away clearly, and up to 200 miles in less detail.

The three Dopplers now in service have impressed weathermen. Forecasters tracking a massive January storm that slammed into the Maryland and Delaware coasts with hurricane force were able to warn people in the Washington area of high winds and heavy rains eight hours in advance.

"Before, the best we could have given was an hour's notice," says James Belville, the head meteorologist at Sterling. "This was the first time a storm of this type was ever captured by radar, and needless to say the data are in great demand now."

In Oklahoma, the Doppler is pinpointing tornadoes more quickly and accurately, and reducing false alarms. "The lead time for tornado warnings has averaged nineteen minutes," Dennis McCarthy, Norman's head meteorologist, tells National Geographic.

Without NEXRAD, the National Weather Service's average warning is about six minutes between the time a twister touches down and slashes into an area.

The national average of tornado false alarms runs about 60 percent, but at Norman it has dropped below 35 percent. Friday predicts that with the new radar, "we're not going to be crying wolf as much as before."

Although the radar has proved accurate in estimating the amount of rain that will pummel an area, weather experts still aren't sure how it will do with snow.

But they're optimistic. After a January snowstorm in the Oklahoma City area, McCarthy said the radar "pinpointed very accurately which areas would receive moderate and light amounts of snow." Based on the data he was receiving, he decided not to let some of his staff leave early. There was some muttering, but the radar correctly showed that the snow would end earlier than expected.

Few weathermen take snow as seriously as does Belville at the Sterling facility. "Every time I put out a warning that closes down the federal work force, it costs the government $43 million," he says. "That puts the forecaster in a pretty tight spot."

Using the new radar to track thunderstorms in Florida, weathermen made a surprising discovery last November and December. They saw small storms moving off the coast. As soon as the storms got to the edge of the warm water over the Gulf Stream, they gave birth to rotating columns of air, which sometimes preceded tornadoes.

CHAPTER 12: MAJOR EQUIPMENT

"We never saw that before," says Friday. "Now we have a strong indication that there is vigorous interaction between thermal patterns generated by the Gulf of Mexico and the weather moving over it."

—By Donald J. Frederick.
Reprinted with permission from the National Geographic Society News Service.

The Weather Channel Gets the Word Out—on TV and through a Variety of New Media Outlets

Even the most accurate high-tech weather warnings will be of little value if they cannot be disseminated widely and rapidly. The information that NEXRAD and other modern weather systems generates needs an outlet channel that nearly everyone can access twenty-four hours a day. The country has that with The Weather Channel, which has one of the largest viewing audiences of any network in the cable television industry, reaching 98 percent of all cable homes.

In addition to a national and local weather service on television through cable and satellite distribution, The Weather Channel offers weather information in a large variety of ways including the Internet (*www.weather.com*), through The Weather Channel Radio Network, on newspaper weather pages with its audiotex service, 1–900–WEATHER. Many of these initiatives are new in the last five years. Here's a story that celebrates The Weather Channel's first ten years and tells how it grew to be a major player in both the cable TV and weather industries.

The Weather Channel Turns 10: 50 Million Subscribers Can't Be Wrong!

Weathercaster Jeff Morrow knew that stalking Hurricane Bob was risky business. The Weather Channel in Atlanta didn't want its on-camera meteorologist or his crew in harm's way. "And yet," Morrow says, recalling the quest for Bob, "we wanted to be there to get the story."

In August of 1991, armed with only a microphone and earphones, he got close enough to the hurricane that he "squinched"—on camera, live, pancake makeup and all. In that moment Bob reminded him of the hazards of standing in the path of a Class 3 hurricane.

Like other storm chasers, Jeff is fascinated by violent storms. As a boy, he was awed by severe weather. He would watch in wonder as wind-driven snows buried Pittsburgh's streets. When summer storm clouds gathered, he would "watch and hope that we would get some severe thunderstorms." Fortunately, he was able to view the lightning strikes of his youth from a safe distance.

That August, however, Morrow deliberately reduced his safety zone. He and his on-site crew in Atlanta got the assignment to "meet" Bob, whose 115-mph winds were then whipping up the Atlantic off the Carolinas. They calculated the storm's probable path, then rushed to get ahead of it, arriving at Kill Devil Hills on North Carolina's Outer Banks toward midnight on August 18. Producer Simon Ross Temperton, assistant producer Lisa Little, and cameraman Norris Lanius (packing a $45,000 minicam) set up the truck to begin broadcasting via satellite the next morning.

These were the waning days of summer vacation and the Outer Banks hotels were packed. "There was a hurricane coming up the coast but people didn't seem to be taking it seriously," Morrow remembers. "It was Saturday night and a lot of people were partying and kind of kidding us."

"There's The Weather Channel. Do you really think Bob's gonna come here?"

"Yeah," we said, "We think it's gonna come right over this area by late tomorrow."

"Hurricane? What hurricane?" vacationers were still asking the next morning in the calm before the storm.

"You guys are doing live updates and it just doesn't look that bad."

True, Morrow concedes, "The wind wasn't that strong and the surf wasn't that high. There were people who came out and actually started to lounge on the beach."

"A little later in the morning, the first outer band of Bob's rain squalls came through. It had some lightning in it. That scared just about everybody out of the hotel. By about 1 or 2 in the afternoon everyone was gone."

Morrow and his crew were still there in the early evening when the next squall scared even them. They were outside on a platform overlooking the beach. Morrow had his back to the storm and was reporting live to The Weather Channel.

"Just as we were going on the air, a lightning bolt hit the ocean right behind me," Morrow said. "I kind of squinched. I saw the flash. There was the lightning and then a big bang."

"Whoa, that was awfully close," Morrow remembers saying.

"Jeff, we saw that flash. Do you think it's safe to be out there?" asked Keith Westerlage, Weather Channel anchor in the Atlanta studio.

"What do you think?—Cut this one off?" Morrow asked producer Temperton.

"Yeah," said Temperton, shaking his head. "We have to get out of here."

Such on-the-spot severe-storm coverage by specially trained meteorologists is now one of the important services television's only 24-hour weather broadcasting operation now offers its many viewers.

Does Anybody Care? Is Anybody There?

But do television viewers care enough about the weather to make all this effort worthwhile? It's a question that was asked frequently after The Weather Channel's launch in 1982, especially when red ink rained down during its first two years on the air. By its 10th anniversary in 1992, network executives could answer that question with a resounding "Yes!" The numbers quieted the skeptics. By 1991, the network that had launched with 2.2 million households had passed the impressive milestone of 50 million subscribers.

The network had also been awarded the cable television industry's highest honor. For its coverage of Hurricane Hugo in September 1989—tropical cyclone expert John Hope tracked the storm from the coast of Africa before it had a name, and meteorologist Dennis Smith met it at Charleston—The Weather Channel earned The Golden Ace Award.

The Weather Channel appears to have reached the status of an American institution. At the American Meteorological Society convention in Atlanta in 1992, the network signed an agreement with the National Weather Service designed to improve dissemination of life-saving weather information, develop new weather services, and create more educational materials like The Weather Channel's documentaries (see below).

"This agreement establishes a framework within which a public agency and a private enterprise can work as partners to benefit the general public," said U.S. Commerce Undersecretary Dr. John A. Knauss, administrator of the National Oceanic and Atmospheric Administration (NOAA).

The Weather Channel's chief executive officer was all sunshine as he and Knauss signed the agreement at a press conference. "NOAA and The Weather Channel each fill unique and important roles in keeping the public informed about the weather, and we have complemented each other for many years," said Michael J. Eckert.

National Weather Service head Joe Friday cited The Weather Channel's unique capacity to continuously translate and communicate weather data gathered by the government's vast weather observation, radar and satellite systems. These government systems are currently undergoing a $3 billion modernization, but the money is earmarked for a new generation of technology, not for weather information dissemination to the public. Friday likened his agency's role to that of a

CHAPTER 12: MAJOR EQUIPMENT

"wholesale" distributor and the role of the media, including The Weather Channel, to that of a "retail" operation.

The Dark before the Dawn

As The Weather Channel begins its second decade on the air, forecasts indicate fair skies ahead. The dark warnings and dire predictions of the early days have passed. "Some had actually projected the exact date we would close in the summer of 1983," says Eckert. In his office northwest of Atlanta, the chief executive recalls why, like the weather, his network's success proved so hard to predict.

One big problem was the revenue expected from advertising, Eckert recalls. When ABC-TV's "Good Morning America" weatherman John Coleman launched his dream of a weather channel, expectations for the whole cable television industry were generally high.

Eckert, hired by Coleman to sell advertising in the West, already had client contacts and saw big bucks on the horizon. "There was great euphoria," Eckert recalls. "From Michigan Avenue (Chicago) to Madison Avenue, advertisers said they would support cable networks. They were interested in the new technology, in getting into homes, and in niche opportunities."

Advertisers also believed cable television would give them leverage with the Big Three networks. By threatening to shift advertising dollars to cable, advertisers saw an opportunity to negotiate lower network rates. Amid this euphoria, Eckert was present at the cable television convention in Las Vegas on May 2, 1982, when the throwing of a ceremonial switch officially launched The Weather Channel.

Advertising sales, however, did not follow as forecast. Part of the reason was the recession of 1981-82, which discouraged investment in the new medium. Then Eckert ran into a stone wall trying to convince clients of the impact of ads on The Weather Channel. "Of the 70 million U.S. television households, we were in one to three million," he recalls. But he couldn't prove an audience of even one million, because the Nielsen firm required a minimum of 13 million homes before it would give a rating. It was a vicious cycle: Eckert had to sell advertising to pay the bills so the station could survive to get the viewers and the ratings.

"You don't know if anyone is watching?" one cereal company executive asked incredulously. As Eckert recalls, the executive "politely asked me to leave his office."

That was the income problem. On the outgo side, startup costs exceeded expectations. The bottom line was that when Coleman left in 1984, the owner of The Weather Channel, Landmark Communications of Virginia, was reporting losses of $800,000 a month. That was the low point—but there was never a funeral.

Viewer Validation

Eckert became vice president of advertising in 1984 and over the next three years witnessed a dramatic reversal of fortune. Fifty of the largest owners of cable systems, who had been getting The Weather Channel free, began paying a fee per subscriber. That fee provided several million dollars to supplement advertising revenues.

This occurred, Eckert explains, because the big cable companies realized that their viewers liked what they'd seen so far. "There was general acceptance," Eckert says, "that The Weather Channel was a viable vehicle. It was serving a useful purpose."

Next, the network invested in viewer research, which showed it should serve different lifestyles. So The Weather Channel began tailoring forecasts for housewives, construction workers, truck drivers, business travelers, farmers, private pilots, and sports fans—including bookmakers who wanted to know, for example, if it was going to rain at RFK Stadium in Washington, D.C., during the NFL playoff games.

Subscribers were now signing up at the rate of five to six million homes per year. That in turn boosted advertising income. "We started to get the Nielsen rating and then we could prove we had a large advertising base," Eckert explains.

Video Actualities

With increasing revenue, network executives decided to improve their weather maps and graphics and expand coverage. Research studies had pointed to the viewer appeal of video "actualities"—footage of some of the more newsworthy weather events from across the country, often relayed from Conus, the Minneapolis-based video network. Through the Conus connection, The Weather Channel also feeds its video reports to TV stations around the country.

Carrying the "actuality" idea a step further, the network created an on-site mobile unit. Staff meteorologist Dennis Smith was dispatched out into the field to meet Hurricane Hugo in 1989. And, on April 15, 1991, Smith and a Weather Channel crew arrived on the scene just minutes after an F5 tornado (261 to 318 m.p.h. winds) killed 17 people at Andover, Kansas, near Wichita. Their broadcast that Friday evening, while other film crews were still arriving, surprised a local TV newsman, who interviewed them on camera Saturday morning.

"We saw your broadcast last night," said the newsman. "When did you all arrive in town?"

"We got here Friday morning," Smith said.

"Wait a minute. The tornado didn't hit until evening. How did you know?"

As Smith explained it, the decision to go to Kansas was "an extremely good guess." The weather situation indicated the likelihood of a serious tornado outbreak "somewhere in northern Oklahoma or Kansas. The most central location was Wichita. It was a big dartboard and we got lucky."

Smith was the logical choice for the assignment. He had been the weatherman for Wichita's KARD-TV (now KSNW-TV) before joining The Weather Channel at its startup, so he knew the town. He also grew up in the region, so severe thunderstorms and tornadoes are his specialty. Besides, he's a weather zealot. Back in Atlanta, he's one of the guys with his "nose pressed against the glass" whenever there's violent weather outside. "We want to be there!" he says.

By 4:30 that Friday afternoon, the crew was set to broadcast live from a balcony on the 26th floor of a Wichita hotel. Two hours later, while they were on the air giving a live update, a severe thunderstorm struck. "We actually saw a little "rope" tornado. Sirens behind us were blaring and we were getting pelted with pea-sized hail. The big killer tornado was on the ground on the backside of the storm, just nine miles away. We immediately went out to the scene."

Two people were killed when the tornado overtook their car. Thirteen more died in a trailer park, despite mobile police warnings—which were videotaped.

How did Smith's broadcast differ from the other TV coverage? "We were able," he says, "to explain a little bit more about the damage viewers were seeing and what caused it. Why you should seek the lowest shelter, . . . some safety rules. The tornado sucked up everything . . . hymnal books were found 100 miles away; canceled checks were found in southern Iowa. There was tremendous destruction."

Later that night, when Smith entered a restaurant he was shocked to see a sign that read: *Temporary Morgue*.

The on-site crew took special precautions in approaching Hurricane Hugo. Hugo was a Class 4 hurricane with winds over 130 mph, the worst to hit South Carolina this century. Smith, fellow meteorologist Cindy Pressler (now a weathercaster for a Lexington, Kentucky, TV station), and their producers and technical staff were getting continuous advice from John Hope in the Atlanta office. Few knew more about what Hugo could do. Hope had been the National Weather Service's senior hurricane specialist when he joined The Weather Channel in 1982.

CHAPTER 12: MAJOR EQUIPMENT

"They didn't quite get to downtown Charleston until after Hugo hit," says Hope, who stayed on the air 17 straight hours on September 21, 1989. "We were a little apprehensive. Had that hurricane gone ashore about 20 to 25 miles down the coast from where it did, there probably would have been numerous fatalities in Charleston.

"I can remember how I felt that night. You know, meteorologists sometimes get wild during a major hurricane. It's such an intense thing. There's a lot of adrenalin pumping. I was also a little depressed, though, because I knew what that hurricane was going to do and I thought a lot of people in that place did not know.

"We followed Hugo from the coast of Africa. It began as a tropical depression off Senegal, southwest of the Cape Verde Islands. We thought it would develop into a hurricane, a very powerful hurricane."

As Hugo neared the U.S. coast, Hope emphasized its dreadful power and the risks. "I knew there was going to be a very high storm surge, 20 feet or so, that would rake the lowlands, do a lot of damage."

Emergency management people who had to make evacuation decisions were watching, Hope says. "They got 100 percent evacuation of the barrier islands, which was very difficult to do. I think the fact that we emphasized the very serious nature of the hurricane as it was approaching South Carolina played a role in saving lives. Who can say for sure? A lot of people were watching us. I went into a hotel in Charleston the next summer and the bellboy told me, 'We're glad to have you in this hotel. You saved a lot of lives in this city.'"

The Weather Channel's role was communication and interpretation. Aside from the on-site team's reports, the channel's forecasters relied on information provided by the National Hurricane Center in Miami. "We just relay their information to the public," Hope explains. "The people at the Hurricane Center will tell you that they are almost totally dependent on the media to get their information out."

—By Leonard Ray Teel. Reprinted with permission from *Weatherwise*.
Edited for the second edition by Connie Malko, Senior Publicist, The Weather Channel

Today's Most Advanced Technology

Long known for its reliable high-tech weather reporting system, The Weather Channel is pioneering new technology for disseminating weather information—both in on-air presentation of its 24-hour cable television weather service and in developing new high-tech outlets for providing weather imformation to consumers
—wherever they are and whatever they are doing. Unlike the local TV weather forecaster who relies largely on news services for weather forecasts, The Weather Channel employs more than 67 meteorologists and spends millions in specialized equipment to produce local reports and forecasts for the entire nation. Using information gathered by the National Weather Service from about 1,800 reporting sites and other sources of raw data, it conducts its own analyses. A new custom-designed computer system integrates all available weather information, data, and advanced meteorological applications into a network of work stations, hastening the speed with which the staff can produce and update its forecasts.

In January of 1997, The Weather Channel installed innovative equipment for enhancing the overall graphics quality of The Weather Channel, a revolutionary development that coincided with the network's move to a new facility and upgrade to an all-digital file server for all graphics. The hardware was created by Silicon Graphics, Inc., "recognized around the world as the absolute premier workstation environment for graphics," noted Mark McKeen, senior vice president of operations for The Weather Channel. The new programming elements include high-resolution 3-D imaging, a broader spectrum of color, clearer fonts, and full animation. The equipment also gives The Weather Channel the capability for a "virtual weather system," which will bring viewers compelling photo-realistic images of real-time weather patterns and even a visual forecast of what the upcoming weather will look like.

The Weather Channel transmits thousands of customized forecasts simultaneously using patented technology called THE WEATHER STAR (which stands for Satellite Transponder Addressable Receiver). Custom equipment at each individual cable system decodes the transmission to extract the one forecast that will give viewers the current information about their own immediate area. Silicon Graphics developed new technology for the transmission of the local forecasts—the Weather Star XL, which opens the way for revolutionary advances in the graphics and features in local weather presentation, scheduled six times every hour on The Weather Channel. The new eight-story facility in Atlanta for The Weather Channel also houses the growing staff and technology that supports the new media ventures of the company. These include: an interactive telephone service (1–900–WEATHER), on-line and multimedia products, and a weather information service for nearly 200 radio stations throughout the country.. The Weather Channel, the country's pre-eminent weather information provider on television, is quickly becoming the primary online source of weather as well. Its popular Web site provides local weather for 1,600 cities worldwide, ski conditions, in-depth storm coverage, airport and flight delays, and local area radar and interactive city maps. Usage soars during severe weather resulting, for example, in 20 million hits in one day during a recent winter storm period. The Internet home page, which was redesigned in February of 1997, is reached at *www.weather.com*. On-line service is also available on CompuServe by typing Go: TWCForum.

"We always believed the network would succeed on the merit of our service and technology," said Chuck Herring, vice president of meteorology operations. "So far, we've managed to maintain a mix that meets the needs of American viewers, and our products and services are used daily in tens of millions of homes," he said.

—Courtesy of Connie Malko, Senior Publicist, *The Weather Channel*

Organizations for Professional Meteorologists

If all this talk of the "big iron" world of professional weather equipment has you daydreaming of a new career goal, we suggest you contact a professional meteorological organization. Write for information. It may be your first step toward an exciting, rewarding career in a field that always will be one of the hottest topics and never will be replaced or become outdated.

American Meteorological Society
45 Beacon Street
Boston, MA 02108
(617) 227-2425

National Weather Association
4400 Stamp Road, Room 404
Temple Hills, MD 20748
(301) 899-3784

National Council for Industrial Meteorologists
600 E. Evans Avenue
Denver, CO 80222

Chapter 12: Major Equipment

When you're planning a trip to the seashore, you'll most likely want the latest official beach forecast. Of course, finding what you want won't be hard, either, because the modern equipment used by today's professional forecasters delivers top-notch accuracy.

Not all of this new, high-tech weather equipment, however, is designed for making sure you enjoy the perfect beach vacation this weekend. Believe it or not, even in our instant, throwaway society, some government projects are aimed at long-term weather research. We learned about one unique and amazing piece of weather-observation equipment whose data output will help make sure that you always can enjoy perfect beach weekends.

Monster Crabs on the Beach

Standing at the end of a 1,840-foot-long pier, oceanographer Charles Long shouts above the tumult of a blustery northeastern: "Strange, violent things happen in the surf zone, especially on days like this."

The 6-foot breakers pounding the Carolina coast, roiling the Atlantic water to a white froth, can be accompanied by wandering sand bars, weird currents and mysterious surges called infragravity waves.

"It's only in the last few years that we've begun to understand some of the things that occur in the surf zone where sand, water and wind meet," says William A. Birkemeier, who heads the Army Corps of Engineers research center here on the north end of the ever-shifting Outer Banks.

The information coming out of Duck [North Carolina] may save millions of dollars as many maritime nations battle damaging beach erosion.

In the United States alone, it is estimated, since 1962 the Army Corps of Engineers has replenished more than 400 miles of beaches with 1.7 billion cubic yards of sand at a cost of about $8 billion.

"Eventually we hope to use the wave and current information that we're getting to predict just how much dune or beach erosion might take place during a given set of circumstances, such as a big storm," says Charles L. Vincent, a scientist at the Coastal Engineering Research Center in Vicksburg, Miss. "The information will also enable us to take more effective countermeasures."

Attracted by the excellent facilities, weather and ocean experts from all over the world flow in and out of Duck.

The reinforced concrete pier, fitted with instruments above and below the waterline, serves as a research platform from which waves, currents, water levels and bottom elevations can be accurately measured, even during the horrendous storms that sometimes batter the area.

But the most unusual research tool is an ungainly contraption with an instrument platform mounted on a 34-foot-high wheeled tripod. Self-propelled on waterfilled tires, the "Coastal Research Amphibious Buggy," or the "crab," can wade into water as deep as 30 feet to take detailed measurements of the ocean bottom.

The visible beach constitutes only a fraction of the actively changing near-shore zone, which near Duck extends seaward about 3,300 feet to a depth of about 26 feet.

The weather has a major impact on all beach systems. Wild and windy conditions increase the number of potentially destructive infragravity waves, little understood until recently. And measurements made by the crab and other instruments have shown that large sandbars can wander seaward as much as 100 feet a day during violent storms.

"They form and get their energy from the normal waves and sea swells that you see," Robert T. Guza, an oceanographer at Scripps Institution of Oceanography in La Jolla, California, tells National Geographic. "Somehow they extract energy from them."

On a calm day, infragravity waves are scarce. Barely perceptible to the naked eye, they may occur a few minutes apart, their small crests separated by a mile or more.

But during storms they proliferate and can be 7 feet high, adding punch to the normal waves battering the beach. Regular waves ride them to shore, like surfers.

"Only recently have we shown that they get real big in storms, causing a lot of water motion and erosion," says Guze.

High-water marks tell beachgoers where such waves have come ashore. "Or if you're standing high and dry in the sand and all of a sudden you're ankle-deep in water, that's a tip-off that an infragravity wave hit your area," Birkemeier says.

Other large waves hitting the surf zone at an angle generate currents along the shore. These currents sometimes become unstable and oscillate like snakes or twisting rivers.

"They can strike with tremendous force," says Joan Oltman-Shay, a Seattle-based scientist who has studied them at Duck.

"You can be standing in the surf zone experiencing a current of about 8 inches a second, which is not enough to kick up a rooster tail, and then four minutes later be knocked downstream with a 5-feet-per-second current," she says.

"It's a horizontal motion, horizontal swirls that are propagating along the shore. If you're standing in one place, you'll pick up the crest of a swirl, the forward motion of a swirl as it moves past you, and the next one's backward motion. It's a tremendously strong signal of velocity."

Daily video images from a 119-foot observation tower show that the pattern of breaking waves in the turbulent near-shore zone changes every year.

The images, combined with data from the crab, are giving scientists the first detailed look at roaming sandbars and other changes in the ocean bottom.

—By Donald J. Frederick.
Reprinted with permission from the National Geographic Society News Service.

The Future of the National Weather Service

Automated weather observation equipment not only is creeping along the bottom of the ocean in North Carolina, it also is being installed all over the country. We would like to close this chapter with a summary of the National Weather Service's current status and a look at its expected future.

The National Weather Service (NWS) states its latest objectives as: "modernizing the NWS through the deployment of proven observational, information processing and communications technologies, and establishing an associated cost effective operational structure. The modernization and associated restructuring of NWS shall assure that the major advances which have been made in our ability to observe and understand the atmosphere are applied to the practical problems of providing weather and hydrologic services to the Nation."

Perhaps their most ambitious modernization program is the Automated Surface Observing System (ASOS). A total of 172 ASOS systems now have been installed at sites across the nation and on the average another system is added almost every day. One of the major benefits of ASOS will be a significant increase in the number of routine observing sites, including many airports where no observation now is available. Another is a nearly continuous "weather watch" at all sites, and extensions to 24 hour coverage at part-time sites.

It is natural and understandable that direct users of manual surface observations, weather observers and other interested persons, tend to assess ASOS in comparison to the venerable manual observation as a standard. This is a tough standard, indeed. It is even tougher when the most complete and detailed observation made by the most skilled and experienced observer is used, and when strict adherence to every rule and guideline in the equally venerable Federal Meteorological Handbook No. 1 is assumed. Unfortunately, our current manual observing program cannot consistently meet that tougher standard, and it is very demanding of staff time and attention. More importantly, the demands on the National Weather Service and other agencies for more observations, with higher resolutions in space and time, and improved services simply cannot be met using current methods.

More than a decade ago Federal agencies undertook a comprehensive re-examination of mission priorities and how best to meet them within practical limitations of cost and technical feasibility. An analogous re-examination has been proceeding in private industries, business and other institutions across the nation and the world. A common result for many, including the National Weather Service, is that a significantly different "way of doing business" is the best, often the only, real choice.

Typically, automation of some labor-intensive functions is an important element, allocation of scarce human resources to the most critical functions that demand human capabilities is a watchword. Whole "patterns" of activity, not just individual functions, must be considered and changed. Effective use of automation seldom is achieved through detailed replication of manual functions.

The appropriate context for understanding and assessing ASOS is the comprehensive modernization and restructuring of the National Weather Service and the National Airspace System. ASOS is a part of a changed pattern of activities and services that must be considered holistically. Many changes are occurring that affect and complement each other. Current and growing needs for observational data and derived information will be met in new and different ways, using different sensors, different systems, different algorithms, different syntheses.

The complexity of the changes is extremely challenging even for those who have worked toward them for more than a decade, day in and day out. It is easy to understand why they are confusing and threatening. Furthermore, we recognize that additional details remain to be worked out, and unforeseen problems and issues are likely to arise. But change is unavoidable.

We are working with the Association of American Weather Observers (AAWO) in maintaining and extending observations of important variables that can't yet be automated or replaced with other information. We agree, for example, that snow depth and accumulation rates and weather equivalents are important and needed by many users including the NWS. We still believe that spotters are an important part of the warning program to save lives.

We are not abandoning manual observing functions that provide information agreed to be essential, still alternative ways of providing the information are available and demonstrated. We are dropping some manual observing elements that were easy to add when a human was "out there anyway," even though someone occasionally uses that data. We are forcing a shift toward a slightly different meaning of familiar variables such as visibility and ceiling. Our belief is that such changes will not reduce aircraft safety, cost lives nor reduce operational efficiencies. But we are listening, adapting and perfecting.

—From an excerpt of a letter by Elbert W. Friday, Jr.,
assistant administrator for Weather Services, National Weather Service.
Reprinted with permission from the *American Weather Observer*.

Summary

Weather information produced by "The Crab," NEXRAD, ASOS, and other such high-tech weather observers no doubt will further lessen the economic impact and social disruptions that raging weather inflicts upon humankind. We'll never truly conquer the weather, but more than ever before we are able to contain its effects on us.

You cannot expect to buy any of the specialized and enormously expensive weather equipment used today by large professional organizations. Still, through outlets such as The Weather Channel, computer bulletin-board systems, and instant weather faxing services, you can have instant access to the valuable information that such major equipment produces in your own home. It's the next best thing to owning it; considering the price tags, it's even better.

This wealth of weather information is readily available in so many forms that we've dedicated the entire next chapter to how and where you can get the latest weather forecasts from nearly anywhere in the world.

CHAPTER 12: MAJOR EQUIPMENT

Products

Ultimeter 2000
Wind & Weather
P.O. Box 2320
Mendocino, CA 95460
(800) 922–9463

The Ultimeter 2000 system offers much of the weather gathering capability of some of the best major equipment, but at some of the lowest prices we've found for a home weather station. It features a keyboard/display unity that shows barometric pressure, indoor and outdoor temperatures, wind speed and direction, and wind chill. An exclusive feature of the Ultimeter 2000 is it triple high–low memory system that displays today's, yesterday's, and long-term minimum and maximum for each weather element, with time and date of the occurrence.

The basic package includes sensors, desk stand, junction box, cables, AC power adapter, and is priced at $379. By adding the optional rain gauge ($90) and outdoor humidity sensor ($110), you'll have a complete weather station. You can even add the Data Logger software ($69) and load the Ultimeter's information directly into a personal computer.

Sensor Weather Data Logger WDL-4
Sensor Instruments Company, Inc.
41 Terrill Park
Concord, NH 03301
(800) 633–1033

This is as close to major weather equipment that most people will ever have in a home. While it is expensive (about $2500 for the full package), it is in the range where serious weather hobbyists can consider installing one at home. If the price seems high, just consider what this much electronic weather equipment would have cost a few years ago! Or, where you would have put if you could afford it!

Naturally, you get absolutely accurate, professional-quality measurements and readouts. But you also get an amazing array of historical data. The WDL-4 will store hourly readings of temperature, barometric pressure, wind speed, precipitation, and humidity for 35 days! And those readings are likely to be lost, either, because the WDL-4 has a battery backup that can last for a month. Of course all data can be loaded directly into a connected personal computer, or transferred over a phone line to a modem-equipped computer. It may be a lot of money for many home weather enthusiasts, but owning the WDL-4 would put you in the class of the professionals.

Chapter 13
How to Get Weather Forecasts and Information

The computer and communications revolution of the past decade has made a plethora of forecasting services available to nearly everyone in the world. In chapter 12, "Major Equipment," you learned that you no longer need to miss sleep to watch the weather on the late news or keep an alert ear to a radio. If you want to know the latest forecast, you can get it within minutes of its becoming available; and you'll know it's been generated by incredible high-tech weather equipment.

All of the popular personal computer bulletin-board systems (BBS) offer weather forecast and reporting systems. Membership in these services is inexpensive and usually provides unlimited access to both local and global weather forecasts for a low, flat monthly fee. Some include color weather maps that can be displayed on your computer screen. Most also provide the latest radar and satellite photos for downloading.

In addition to professionally prepared weather forecasts, there are services that provide information that you can use to prepare your own forecasts. Those services are covered in chapter 11, "How to Become a Weather Forecaster." This chapter deals more with getting completed weather forecasts and compiled weather-related data. Using such data, you can compare the results of the professionals with your own forecasting results.

Another example of how to use downloaded weather data is with computerized hurricane-tracking software, reviewed later in this chapter, that can keep you up to date with an accurate track of a developing hurricane and help you be better prepared for its arrival. Downloaded data also can help you conduct detailed studies of exceptional weather events.

If you don't own a personal computer yet, you don't need to feel left out. Before we list sources for PC-based weather data and forecasts, we'll list sources for weather forecasts available to everyone. No special equipment is needed to "download" these trusty, inexpensive weather forecasts—and they're "bug-free."

NOAA Weather Radio Network

The National Oceanic and Atmospheric Administration broadcasts continuous weather reports directly from the offices of the National Weather Service. Broadcasts are made on one of seven high-band FM frequencies ranging from 162.40 to 162.55 megahertz. A number of manufacturers produce special weather radios that operate on these frequencies; in addition, many standard radios offer the "weather band" as well as the usual AM/FM frequencies.

NOAA Weather Radio Frequencies

Legend

Frequencies are identified as follows:
1–162.550 MHz
2–162.400 MHz
3–162.475 MHz
4–162.425 MHz
5–162.450 MHz
6–162.500 MHz
7–162.525 MHz

Location	Frequency
Alabama	
Anniston	3
Birmingham	1
Demopolis	3
Dozier	1
Florence	3
Huntsville	2
Louisville	3
Mobile	1
Montgomery	2
Tuscaloosa	2
Alaska	
Anchorage	1
Cordova	1
Fairbanks	1
Homer	2
Juneau	1
Ketchikan	1
Kodiak	1
Nome	1
Haines	1
Seward	1
Sitka	1
Valdez	1
Wrangell	2
Yakutat	2
Arizona	
Flagstaff	2
Lake Powell	1
Phoenix	1
Tucson	2
Yuma	1
Arkansas	
Fayetteville	3
Fort Smith	2
Gurdon	3
Jonesboro	1
Little Rock	1
Mountain View	2
Star-City	2
Texarkana	1
California	
Bakersfield	1
Coachella	2
Eureka	2
Fresno	2
Los Angeles	1
Lindsay	6
Monterey	2
Point Arena	1
Redding	1
Sacramento	1
San Diego	2
San Francisco	2
San Luis Obispo	1
Santa Barbara	2
Colorado	
Alamosa	3
Colorado Springs	3
Denver	1
Fort Collins	5
Grand Junction	1
Greeley	2
Longmont	1
Pueblo	2
Sterling	2
Connecticut	
Hartford	3
Meriden	2
New London	1
Delaware	
Lewes	1
District of Columbia	
Washington, D.C.	1
Florida	
Bell Glade	2
Daytona Beach	2
Fort Myers	3
Gainesville	3
Jacksonville	1
Key West	2
Melbourne	1
Miami	1
Orlando	3
Panama City	1
Pensacola	2
Tallahassee	2
Tampa	1
West Palm Beach	3
Georgia	
Athens	2
Atlanta	1
Augusta	1
Baxley	7
Chatsworth	2
Columbus	2
Macon	3
Pelham	1
Savannah	2
Valdosta	6
Waycross	3
Waynesboro	4
Hawaii	
Hilo	1
Honolulu	1
Kokee	2
Mt. Haleakala	2
Waimanalo	2
Idaho	
Boise	1
Lewiston	1
Pocatello	1
Twin Falls	2
Illinois	
Champaign	1
Chicago	1
Marion	4
Moline	1
Peoria	3
Rockford	3
Springfield	2
Indiana	
Bloomington	5
Evansville	1
Fort Wayne	1
Indianapolis	1
Lafayette	3
Marion	5
South Bend	2
Terre Haute	2
Iowa	
Cedar Rapids	3
Des Moines	1
Dubuque	2
Sioux City	3
Waterloo	1
Kansas	
Chanute	2
Colby	3
Concordia	1
Dodge City	3
Ellsworth	2
Topeka	3
Wichita	1
Kentucky	
Ashland	1
Bowling Green	2
Covington	1
Elizabethtown	2
Hazard	3
Lexington	2
Louisville	3
Mayfield	3
Pikeville	2
Somerset	1
Louisiana	
Alexandria	3
Baton Rouge	2
Buras	3
Lafayette	1
Lake Charles	2
Monroe	1
Morgan City	3
New Orleans	1
Shreveport	2
Maine	
Caribou	7
Dresden	3
Ellsworth	2
Portland	1
Maryland	
Baltimore	2
Hagerstown	3
Salisbury	3
Massachusetts	
Boston	3
Hyannis	1
Worcester	1
Michigan	
Alpena	1
Detroit	1
Flint	2
Grand Rapids	1
Houghton	2
Marquette	1
Onondaga	2
Sault Sainte Marie	1
Traverse City	2
Minnesota	
Detroit Lakes	3
Duluth	1
International Falls	1
Mankato	2
Minneapolis	1
Rochester	3
Saint Cloud	3
Thief River Falls	1
Willmar	2
Mississippi	
Ackerman	3
Booneville	1
Bude	1
Columbia	2
Gulfport	2
Hattiesburg	3
Inverness	1
Jackson	2
Meridian	1
Oxford	2
Missouri	
Columbia	2
Camdenton	1
Hannibal	3
Hermitage	5
Joplin/Carthage	1
Kansas City	1

CHAPTER 13: WEATHER FORECASTS

Location	Frequency
St. Joseph	2
St. Louis	1
Sikeston	2
Springfield	2
Montana	
Billings	1
Butte	1
Glasgow	1
Great Falls	1
Havre	2
Helena	2
Kalispell	1
Miles City	2
Missoula	2
Nebraska	
Bassett	3
Grand Island	2
Holdrege	3
Lincoln	3
Merriman	2
Norfolk	1
North Platte	1
Omaha	2
Scottsbluff	1
Nevada	
Elko	1
Ely	2
Las Vegas	1
Reno	1
Winnemucca	2
New Hampshire	
Concord	2
New Jersey	
Atlantic City	2
New Mexico	
Albuquerque	2
Clovis	3
Des Moines	1
Farmington	3
Hobbs	2
Las Cruces	2
Ruidoso	1
Santa Fe	1
New York	
Albany	1
Binghamton	3
Buffalo	1
Elmira	2
Kingston	3
New York City	1
Riverhead	3
Rochester	2
Syracuse	1

Location	Frequency
North Carolina	
Asheville	2
Cape Hatteras	3
Charlotte	3
Fayetteville	3
New Bern	2
Raleigh/Durham	1
Rocky Mount	3
Wilmington	1
Winston-Salem	2
North Dakota	
Bismarck	2
Dickinson	2
Fargo	2
Jamestown	2
Minot	2
Petersburg	2
Williston	2
Ohio	
Akron	2
Cambridge	3
Cleveland	1
Columbus	1
Dayton	3
Lima	2
Sandusky	2
Toledo	1
Oklahoma	
Clinton	3
Enid	3
Lawton	1
McAlester	3
Oklahoma City	2
Tulsa	1
Oregon	
Astoria	2
Brookings	1
Coos Bay	2
Eugene	2
Klamath Falls	1
Medford	2
Newport	1
Pendleton	2
Portland	1
Roseburg	1
Salem	3
Pennsylvania	
Allentown	2
Clearfield	1
Erie	2
Harrisburg	1
Johnstown	2
Philadelphia	3
Pittsburgh	1
State College	3

Location	Frequency
Towanda	3
Wellsboro	1
Wilkes-Barre	1
Williamsport	2
Puerto Rico	
Maricao	1
San Juan	2
Rhode Island	
Providence	2
South Carolina	
Beaufort	3
Charleston	1
Columbia	2
Cross	3
Florence	1
Greenville	1
Myrtle Beach	2
Sumter	3
South Dakota	
Aberdeen	3
Huron	1
Pierre	2
Rapid City	1
Sioux Falls	2
Tennessee	
Bristol	1
Chattanooga	1
Cookeville	2
Jackson	1
Knoxville	3
Memphis	3
Nashville	1
Shelbyville	3
Waverly	2
Texas	
Abilene	2
Amarillo	1
Austin	2
Beaumont	3
Big Spring	3
Brownsville	1
Bryan	1
Corpus Christi	1
Dallas	2
Del Rio	2
El Paso	3
Fort Worth	1
Galveston	1
Houston	2
Laredo	3
Lubbock	2
Lufkin	1
Midland	2
Paris	1
Pharr	2

Location	Frequency
San Angelo	1
San Antonio	1
Sherman	3
Tyler	3
Victoria	2
Waco	3
Wichita Falls	3
Virgin Islands	
St. Thomas	3
Utah	
Logan	2
Cedar City	2
Vernal	2
Salt Lake City	1
Vermont	
Burlington	2
Marlboro	4
Windsor	3
Virginia	
Heathsville	2
Lynchburg	1
Norfolk	1
Richmond	3
Roanoke	2
Washington	
Neah Bay	1
Olympia	3
Seattle	1
Spokane	2
Wenatchee	3
Yakima	1
West Virginia	
Beckley	6
Charleston	2
Clarksburg	1
Gilbert	7
Hinton	4
Moorefield	7
Spencer	6
Sutton	5
Wisconsin	
La Crosse	1
Green Bay	1
Madison	1
Menomonie	2
Milwaukee	2
Park Falls	6
Wausau	3
Wyoming	
Casper	1
Cheyenne	3
Lander	3
Sheridan	3

USA Today

The back page of *USA Today,* published nationwide Mondays through Fridays, has the most widely read weather forecast anywhere. Using colorful graphics, it will give you a quick overview of U.S. weather at a glance. Individual forecasts for major cities are included, covering the forecasts for today and tomorrow. It also provides temperature and expected cloud-cover information for about forty major cities around the world.

The daily forecasts often are accompanied by a short, colorful lesson on a weather phenomenon that is expected to affect some area of the United States that day. For example, an issue in May might include a graphically illustrated article about how thunderstorms spit out hail.

If you are traveling to another city, perhaps you will want more details on your destination weather. The *USA Today* weather page lists a telephone "900" number that will bring you weather forecasts, time, temperature, and travel conditions for some 650 U.S. cities, plus up-to-date currency-exchange rates for many foreign cities. For access, dial (900) 555-5555, then press "11" and the area code of the desired city. For foreign-city forecasts and exchange rates, press the first three letters of the city's name after pressing "11."

Free Telephone Recorded Forecasts

These telephone recordings can provide a quick check of forecast weather for today and tomorrow. But for longer-range forecasts, maps and radar summaries, or for cities outside the United States, you'll need to use one of the commercial PC bulletin-board systems. If you own a PC that is equipped with a modem, you can tie into these weather sources using any telephone line. Charges for these services vary, so check with the sources for the latest cost figures.

Albany, NY	(518) 476-1122
Albuquerque, NM	(505) 243-1371
Atlanta, GA	(404) 936-1111
Birmingham, AL	(205) 942-8430
Bismarck, ND	(701) 223-3700
Boise, ID	(208) 342-8303
Boston, MA	(617) 567-4670
Buffalo, NY	(716) 634-1615
Cape Cod, MA	(617) 771-0500
Caribou, ME	(207) 496-8931
Charleston, WV	(304) 344-9811
Cheyenne, WY	(307) 635-9901
Chicago, IL	(312) 298-1413
Cincinnati, OH	(513) 241-1000
Cleveland, OH	(216) 931-1212
Columbia, SC	(803) 976-8710
Denver, CO	(303) 639-1212
Des Moines, IA	(515) 288-1047
Detroit, MI	(313) 941-7192
Elko, NV	(702) 738-3018
El Paso, TX	(915) 778-9343
Eugene, OR	(503) 484-1200
Fort Worth, TX	(817) 336-4416
Great Falls, MT	(406) 453-5469
Indianapolis, IN	(317) 222-2362
International Falls, MN	(218) 283-4615
Jackson, MS	(601) 936-2121
Jacksonville, FL	(904) 757-3311
Key West, FL	(305) 296-2011
Little Rock, AR	(501) 834-0316
Los Angeles, CA	(213) 554-1212
Louisville, KY	(502) 363-9655
Lubbock, TX	(806) 762-0141
Memphis, TN	(901) 757-6400
Miami, FL	(305) 661-5065
Milwaukee, WI	(414) 744-8000

CHAPTER 13: WEATHER FORECASTS

Minneapolis, MN	(612) 452–2323	Reno, NV	(702) 793–1300
Myrtle Beach, SC	(803) 744–3207	St. Augustine, FL	(904) 252–5575
Nags Head, NC	(919) 995–5610	St. Louis, MO	(314) 928–1198
New Orleans, LA	(504) 465–9212	St. Petersburg, FL	(813) 645–2506
New York, NY	(212) 315–2705	Salt Lake City, UT	(801) 575–7669
Ocean City, MD	(301) 289–3223	San Antonio, TX	(512) 828–3384
Oklahoma City, OK	(405) 360–8106	San Francisco, CA	(415) 936–1212
Omaha, NE	(402) 571–8111	Savannah, GA	(912) 964–1700
Orlando, FL	(305) 851–7510	Seattle, WA	(206) 526–6087
Philadelphia, PA	(215) 627–5578	Sheridan, WY	(307) 672–2345
Phoenix, AZ	(602) 957–8700	Shreveport, LA	(318) 635–7575
Pittsburgh, PA	(412) 644–2881	Sioux Falls, SD	(605) 336–2837
Portland, ME	(207) 775–7781	Topeka, KS	(913) 234–2692
Portland, OR	(503) 236–7575	Tulsa, OK	(918) 477–1000
Raleigh, NC	(919) 860–1234	Virginia Beach, VA	(804) 853–3013
Redding, CA	(916) 221–5613	Washington, D.C.	(202) 936–1212
Rehoboth, MD	(302) 856–7633	Wichita, KS	(316) 942–3102

CompuServe

CompuServe offers a full array of computer-delivered weather forecasts, covering cities all over the world. For only $9.95 per month, users gain unlimited access to CompuServe's weather services. These services are divided into four basic groups so that users can jump quickly to a desired area, requiring a minimum of navigation through menu systems.

For IBM-compatible-computer users, a Windows version of the CompuServe Information Manager (CIM) is a lot of fun to use. The integrated nature of Windows—especially its universal clipboard—enables users to "cut and paste" information between the CIM and all other Windows applications. For example, you can copy forecast information from a CIM weather service to the clipboard, then paste it directly into your word processor. Here's a listing of the four CIM weather services:

- WEATHER. This choice contains various weather products. For U.S. locations, a wide variety of National Weather Service reports and Accu-Weather maps are provided. For locations outside the United States, Accu-Weather maps, current conditions, and three-day forecast reports are available. Weather information also is available as part of some aviation products. (See chapter 10, "Weather for Pilots.")
- WEATHER MAPS. Using CompuServe CIM or any GIF-compatible program, you can display sixteen-color or black-and-white maps of North America, the United Kingdom, continental Europe, the Pacific Rim, and Australia/New Zealand. Each section offers a satellite map, current conditions, and forecast maps. Radar depiction and regional wake-up maps are also available for the United States.
- WEATHER REPORTS. For U.S. locations, the National Weather Service features short-term state and local forecasts, severe-weather alerts, precipitation probability, state summaries, and daily climatological reports.
- U.K. WEATHER. The HMI version of U.K. Weather contains short-term weather reports for

more than a dozen cities in England, Scotland, Wales, and Northern Ireland as well as high- and low-resolution weather maps. Also available are reports and maps for sites in continental Europe, North America, and the Pacific Rim. The HMI version works with DOS CIM 1.3 or higher and with Mac CIM 1.5 or higher.

America-Online

The America Online News and Finance department gives you up-to-the-minute news, weather, and business information. The Today's News and Weather area lets you read the top news stories for the nation and the world. It enables you to get a summary of weather information and download color weather maps. If you use the Windows version, you can use the Windows clipboard to cut and paste weather information and graphic maps from America Online into any other Windows application.

You also can look at the national news in *USA Today*. For those interested in business and finance, there is national and international business news and weather, financial-planning advice, an online real estate service, stock prices, and stockmarket commentary. Contact America Online at (800) 827–6364 for sign-up information.

Computerized Hurricane Tracking

PC Weather Products
45 East Blackland Court
Marietta, GA 30067
(800) 242–4775

As an aid for getting forecasts about hurricane tracks and as a research tool, PC Weather Products of Marietta, Georgia, has released a computerized mapping program and hurricane archive database for storm tracking and analysis. Why do you need a hurricane-tracking program? After all, couldn't you do the same thing with a pencil and a map? Perhaps, but not with the same accuracy and detail.

The program, entitled HURRTRAK, is designed for people who have a keen interest in the latest information about hurricanes. If you use a PC and take storm tracking seriously, you'll appreciate these features:

- Detailed tracking charts for trackers who want to know more than the fact that there is a storm over South Florida. This system uses fifteen detailed, predefined tracking charts that allow the user to see that the eye is over Homestead and the eye wall is over south Miami. There is also a user-defined chart that can be defined for just about any area and size.
- Graphical representation of eye, eye wall, and areas of storm and hurricane-force winds.
- Storm data next to each observation point on the tracking chart.
- Two hundred predefined city locations, with the ability to add as many more as you want.
- A separate hurricane history (HURRHIST) program that lets you do all types of queries against the historical database.

The most recent release of HURRTRAK is Version 6.0. It includes the companion program HURRHIST. HURRTRAK, written by a meteorologist, is very fast, professional, and easy to use. Installation from the double-density 3½-inch diskette went smoothly as the HURRTRAK install program created a directory on my hard disk and then uncompressed the files as it installed them. It's got fourteen VGA resolution maps, historical storm-track data for all hurricanes in the past 106 years, program documentation, and the HURRTRAK and HURRHIST programs compressed onto only one diskette.

Weather-satellite users will feel at home with the HURRTRAK opening screen as they're treated to a close-up satellite image of the eye of a hurricane. When the opening antics are out of the way, the program settles down to business with an easy-to-use menu system and online help/tutorial mode.

After loading a past or current storm, the analysis feature allows the selected storm to be reviewed. The database includes date, time, location, pressure, and wind speed. These data can be printed, but the best way to view them is graphically, on one of the fourteen high-resolution maps. Viewed this way, each data point is shown on the map along with chart notes. The notes represent the barometric pressure and wind speed at each point in the course of the storm. Sometimes the map can get cluttered with chart notes at each point, but you can plot the storm without notes.

Nice features include the city names on the map and the distance the storm is from each city. Television and radio announcers often mention a city near the storm, but that means little unless you are familiar with local geography. An example is the recent Hurricane Andrew, which passed near Lafayette, IntraCoastal City, and Acadia, Louisiana. With HURRTRAK I was able to see where the storm path actually lay and where these cities are located.

When graphically displaying a storm, the program offers several ways to select a map to plot the storm. My favorite is to let the computer pick the best map. You can also graphically choose your sector from a full overview map or use the menu. Another program feature allows you to create a custom tracking map for any area.

You may invoke the highly entertaining Forecast mode and watch HURRTRAK search through more than 100 years of history to compare the current storm with a composite of similar storms. It then plots a forecast track based on historic records. How does it do this? I don't know, but to quote the manual (on the disk): "The forecast track is based on a time sliding weighted average of present movement and 100+ years of climatology (1886–1991)." Nevertheless, the forecast track is fun to see.

This HURRHIST feature, for me, is the highlight of the program. Its database displays a graphical presentation of past storms since 1886. As is typical of any database, you have a number of ways to select storms for analysis: by name, date range, wind-speed ranges, and pressure. Other interesting options allow plotting of pressure versus winds and a comparison of your selected storm to other storms that tracked through your area in the past century.

I wanted to use this to see how many storms passed near my vacation home outside of Wilmington, North Carolina. After a few seconds I had to stop the program because the screen was cluttered with storm tracks—and it still was analyzing the 1800s! On the second attempt I asked only for storms with winds in excess of 100 mph. Once again the screen filled with storm tracks. I had to ask myself why anyone would live here, at ground zero.

As a storm path is plotted, each point is accompanied by a corresponding tone—lower-pitched tones for lesser winds and higher tones as the storm intensifies. You can tell by listening to its sound pattern if a storm was severe or not. I was playing with this feature on my laptop in the living room one afternoon, and the response from others in the room was, "Oooh, which one was that?" "Allen," I replied. We could feel the severity of each storm as the pitch went up and the storm track appeared on screen.

If you use this program, look up my favorite hurricane: 1971's Ginger. This storm looks like a child's scribbling on the screen as it wanders back and forth, up and down the Atlantic. The plot starts with a low tone, rises, lowers, rises again, then lowers as the storm crisscrosses the Atlantic heading west, then east, then north, then south, crossing over its own path several times. It wasn't a serious hurricane, but it had a most interesting path.

Is there anything I don't like about this program? Yes. While the product is beautiful on a color VGA, with appropriate choice of colors for all plots, graphs, and menus, it needs a monochrome or LCD option for laptop users. Although not a necessity, a printed supplement for the manual on disk would be convenient.

You should add both of these programs to your program library because they are a tremendous bargain. The storm history and storm tracking have much educational value. The storm database alone is a valuable research and reference tool. The ability to look up comparative information when a storm is in progress or threatening is informative. Tracking the storms as they take place allows you to see for yourself, on many different maps with many different perspectives, where the storm lies in relation to surrounding geography or in relation to previous storms. These are things you certainly cannot do with a pencil and a map.

Other features are included, but the program is too powerful to detail the whole system here. The price for the consumer Windows version is $69. HURRTRAK Professional Version 7 is available for $249. The PC Weather Products Web Site (www.pcwp.com) includes downloadable upgrades and full information on the latest version of all products, as well as technical support.

—Review by Tom Glembocki

Paramax Weather Information Services

Paramax Weather Information Services
211 Gale Lane
Kennett Square, PA 19348
(215) 444–2400
fax: (215) 444–2420

Paramax Weather Information Services is a commercial provider of standard and custom weather data, weather display systems, and services to meet specific targeted market areas. Its typical customers are distributors of weather information throughout industry and government.

Paramax Weather Information Services launched into the weather-data, -products, and -services business by becoming a NEXRAD Information Dissemination Service (NIDS) provider. The NEXRAD data service highlights its product line, where its delivery, accuracy, availability, and support are well regarded.

The product line includes a weather-product database; PC software products to receive, display, and manipulate (zoom, loop, etc.) weather products; and workstation-based products to receive, display, manipulate, and combine weather-data products interactively. NIDS Training, twenty-four-hour hotline support, engineering services, facility services, and telecommunications services round out their capability as a complete provider of weather-system solutions.

Paramax PC-Based Software Products

Paramax offers IBM-compatible, DOS- and Windows-based receive and display products for radar, satellite, and lightning. The hallmark product is the NEXRAD radar application, tentatively named WeatherView™, which is a full-featured Windows application for selection, receipt, and display of all the NEXRAD NIDS products. WeatherView™ supports unattended auto dial, custom communications port setup and protocol monitor, custom product selection, zoom up to sixteen times magnification, a navigation tool (displaying latitude/longitude and range between user-selected locations), looping with pause and forward/reverse single step, and numerous map-overlay options (state, county, water, rivers/streams, railroads, highways/roads, administrative boundaries, NWS reporting sites, and polar range rings). Products can be saved in .BMP or NEXRAD formats.

Paramax has an (800) hotline that is staffed twenty-four hours per day, seven days per week, with trained technicians and operators. All Paramax products and services are supported. Operators monitor data-source sites, communications and network status, and customer-communications status. Each operator is familiar with all products and has immediate access to engineering and management staffs for problem diagnosis and resolution.

The National Weather Data Repository

The base product is the National Weather Data Repository, named WeatherBase™. The repository is a centralized database containing unaltered weather data from various sensor sources, plus Value Added Products as they are developed. Sensor data include:

- NEXRAD (WSR-88D) Radar
- Conventional Radar (WSR-57 and 74)
- GOES Tap Satellite, including GMS and METEOSAT
- Lightning
- DIFAX, Public Products, Domestic Products, International Products, and FAA 604 Aviation

Products
- NOAA Weather Wire
- Wind Profiler Demonstration Network
- GOES Direct
- Other services are expected

The database also contains national and regional reflectivity mosaics. Additional mosaic products are constantly under development. The proprietary On-Demand Protocol is used on dial-in and dedicated asynchronous or synchronous interfaces at up to 14.4 Kbps. A continuous output option is available for high-volume customers at speeds of 9.6 Kbps to 1.544 Mbps. Both low-volume and high-volume satellite distributions are planned.

Personalized Services

Paramax offers a comprehensive two-day training seminar on the use and interpretations of its NIDS products. Taught by NEXRAD-experienced meteorologists, the seminar covers an overview of Doppler radar, an overview of NEXRAD, an in-depth look at all NIDS weather products, and NEXRAD data applications. Courses are offered throughout the United States. On-site classes also are available.

Paramax engineers can design, implement, test, and install various receive, collection, dissemination, and display systems and local-area networks. These systems are designed based on each customer's specifications and are fully supported by Paramax.

Paramax also provides facility space for customer-dissemination systems that are directly connected to WeatherBase. This includes operator support, routine maintenance, diagnostic assistance, and generator-power backup.

Accu-Weather

Accu-Weather
619 West College Avenue
State College, PA 16801
(814) 234–9601, ext. 400
fax: (814) 238–1339

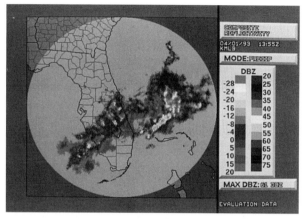

Accu-Weather is the most comprehensive commercial source of weather information available. It offers complete, up-to-the-minute local, national, and worldwide weather data, forecasts, maps, and graphics. The Accu-Data dial-in database delivers all available worldwide weather information twenty-four hours a day. (See the review of access software below.)

Accu-Data is Accu-Weather's state-of-the-art database designed and maintained by its staff of seventy-five professional meteorologists and forty-five computer scientists. They offer a wide variety of products and services backed by thrity years of weather experience. You can select products from more than 35,000 weather-data types, including surface observations, upper-air data, forecasts, watches, warnings, climatological summaries, and marine data. You can download and sort this wealth of information by site, region, state, or country and can specify time periods. Here are some examples:

- **NWS Official Forecasts.** These include local and city forecasts. You can select forecasts by other criteria as well, such as zone, state, marine, three to five days, six to ten days, thirty days, and ninety days.
- **Severe Weather Reports.** You can stay up to date on storm warnings and watches, bulletins and advisories, and flooding and damage reports. You also can participate in general discussions and learn all about severe storms.
- **Aviation and Marine Reports.** Hourly observa-

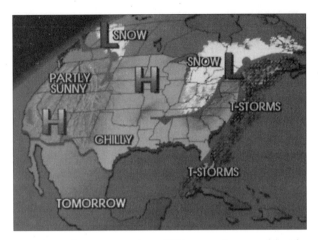

tions from more than 100 countries, worldwide ship and buoy reports, high-resolution soundings, stability indices, freezing levels, route weather briefings and terminal forecasts, area forecasts, winds aloft, SAs, and pilot reports.

- **Weather Maps.** Accu-Weather's exclusive Advanced Map Plotting System (AMPS) enables you to select from more than a million high-resolution maps and weather graphics. Your selections can be sized and contoured to include any area of the world with a few keystrokes. You then can overlay the data you want in nearly any form or format.
- **Color Images.** There are more than 2,000 color-graphics displays available. These are the same ones used by professional meteorologists, such as television weather anchors and government agencies. Products include radar and Accu-Weather's exclusive RadarPlus, current and forecast surface- and upper-air maps, temperature maps, and satellite images. And you can get real-time downloads of all 137 Doppler radars and the hottest new weather product: NEXRAD, in standard and UltraGraphix formats.
- **Lightning Data.** You can see the location, intensity, and frequency of lightning strikes in real time. Set up this service and watch lightning strikes as they occur to help you locate thunderstorms before they appear on radar.
- **DIFAX.** Accu-Data offers instant access to National Weather Service DIFAX images, literally within seconds of becoming available. More than 300 maps and charts are available through DIFAX.
- **Global Weather Services.** Accu-Weather offers worldwide forecasting services, including computerized forecasts up to seven days in advance.
- **Miscellaneous.** Going well beyond mere weather reporting, Accu-Weather offers many other services, including weather-history files, climatological summaries, normals and records, river-stage reports, road-condition reports, sunrise and sunset times, solar data, air-stagnation data and pollution advisories, ski and recreation forecasts and reports, and hydrological and seismic data.

Since the weather never stops, nor takes a holiday, neither does Accu-Weather. It maintains a support staff twenty-four hours a day, 365 days a year to assist you and answer technical questions.

It also has a quarterly newsletter that will keep you abreast of the latest high-tech developments in the weather industry. Regular columns and features help you get more out of the Accu-Data databank, saving you time and connect charges. Accu-Weather also answers questions commonly asked of its support staff and gives you a forum to get your questions answered. The newsletter will make sure you don't miss out on exciting new Accu-Weather products and services.

Teachers will appreciate Accu-Weather's excellent instructional modules, including teacher's guides and student workbooks, for meteorology courses. These beautifully done wire-bound manuals include text written for students by top weather professionals plus a host of charts and graphs to help it all make sense.

Other Weather Forecasting Options

Be sure to see additional listings for forecasting programs and products at the end of chapter 10, "Weather for Pilots."

CHAPTER 13: WEATHER FORECASTS

Accu-Weather Forecaster
Software Toolworks
60 Leveroni Court
Novato, CA 94949
(415) 883–3000
fax: (415) 883–3303

This software application enables you to log directly onto America's premier weather forecasting service, Accu-Weather. It's perfect for checking the weather for vacations, business trips, picnics, farming or gardening, sailing, flying, sports events, construction projects, and open-air concerts. The software lets you predefine your weather request, then log on automatically and quickly download exactly what you want, thus minimizing online fees. You can process this information and view weather maps offline at your leisure.

If you'd prefer to browse, you can search the world for the highest and lowest temperatures or winds or any other weather-data type. Accu-Weather offers a total of more than 35,000 weather products covering the world, so you'll have plenty to look for.

Program purchase price includes one hour of connect time, but it must be used within the first thirty days of initial log on. In addition to online charges, you'll have to pay long-distance charges to the 814 area code in Pennsylvania or use Accu-Weather's 800 number, with a surcharge of 19 cents per minute. Online charges for individuals (rates shown for 2400 baud) are 27 cents per minute between 7:00 P.M. and midnight, 15 cents per minute between midnight and 5:00 A.M., and 59 cents per minute between 5:00 A.M. and 7:00 P.M., and there is a monthly minimum of $9.95. Rates for commercial institutions are approximately double these rates.

Program purchase also includes the Forecasting Guidebook by Professor Eric W. Danielson of the Hartford College for Women. It's written specifically for use with the Accu-Weather Forecaster software and explains how to get forecasts and how to create them using your own data as well as Accu-Weather data.

For more about the weather information available or about rates, see the listing above or call an Accu-Weather marketing representative at (814) 234–9601, extension 400. It requires an IBM-compatible PC, 640K RAM, a hard disk, Hayes-compatible modem (1200 or 2400 baud), and a color-graphics monitor. It is also available for Macintosh. Both versions list for $89.95.

The WeatherCycler
The Weather School
5808 Tudor Lane
Rockville, MD 20852
(301) 230–8985

If you fly, sail, fish, farm, garden, hunt, boat, ski, snowmobile, travel, hike, or camp, you'll love how The WeatherCycler slide chart can keep you abreast of changing weather conditions. You can use it to make sense out of the weather maps and numbers from television, radio, and newspapers. It helps you focus forecasts so you'll understand how they will affect your day's activity. It will also help you forecast the weather yourself when other sources are not available.

Its thin $8\frac{1}{2}$ by 11-inch size enables you to pack it anywhere you go. Since it's priced at only $7.95, you'll be able to take one for everybody in your group. It's also an excellent teaching tool and comes in a teacher's pack, which includes fifteen regular WeatherCyclers, one instructor's Weather-Cycler, thirty sets of study activities, and an instructor's manual for $159.95.

Meteorology Services
Universal Weather & Aviation
8787 Tallyho
Houston, TX 77061
(800) 231–5600
fax: (713) 943–4688

Universal offers a wide range of weather-related information services to offshore marine entities, petrochemical refining operations, public utilities, construction companies, entertainment and information media, professional sports organizations, and emergency-management operations. Its staff of more than fifty meteorologists are located in four U.S. offices. Universal provides forecasts and reports via telephone, fax, telex, SITA, ARINC, AFTN, and a number of radio connections. Subscribers get Universal's weather software, which allows users to obtain immediate color weather graphics (satellite imagery, worldwide surface observations, surface- and upper-level wind conditions), and computerized text on a PC.

Weather Watch

SaySoft
431 Roswell
Indianapolis, IN 46234
(317) 271–3622

This is a weather processor for IBM-compatible PCs, with a complete, state-of-the-art graphical user interface to present nearly every imaginable weather report or forecast. It accepts data from most commercial sources, such as CompuServe or DUATS. It also accepts input directly from home weather-station instruments. It comes with thirty-day money-back guarantee to ensure satisfaction. Order direct from SaySoft for $149.95.

AWOS

Artais Weather-Check
4660 Kenny Road
Columbus, OH 43220
(800) 327–2967
fax: (614) 451–0229

Artais makes mobile weather stations that generate reports in the ICAO and WMO formats and meet the performance requirements of the FAA for Automated Weather Observing Systems (AWOS). The Artais system provides a deployable, "off-the-shelf" weather observation that can be set up in twenty minutes by one person. The units are fully self-contained, automatic, and compact, and are light enough to be air-shipped. They have a built-in generator so they can operate for long periods unattended; and they can be built into a wide variety of custom packages, such as special-use vehicles.

The Artais AWOS-2000 is a state-of-the-art system that monitors, records, and communicates a full range of weather information via VHF/UHF radio in several languages. It also transmits in digital format to a central weather station using UHF or satellite radio links. This is truly professional equipment, in use by the U.S. military services and the FAA.

Weather Satellite Acquisition

Satellite Data Systems, Inc.
P.O. Box 219
Cleveland, OH 56017
(507) 931–4849

SDS enables any IBM-compatible PC to acquire weather data from orbiting satellites. Its ESC-102 WeatherFax satellite card feeds its SDSfax PLUS software so you can "ingest" images in several formats. An Auto Acquire feature permits unattended operation that saves images to disk in the PCX graphics format, which can be loaded into nearly any word processor, paint program, or desktop-publishing software.

A Crosshair Cursor permits zooming in on any location within any image and can overlay latitude/longitude points or gridding. SDS has an extensive line of satellite antennas and offers convenient, "one-stop-shopping" packages that include everything but the PC: ESC-102 PC card, software, receiver, antenna, power supply, preamp/downconverter, cable, and connectors. A basic system for acquiring from polar-orbiting satellites costs $1,435. You can upgrade individual components or buy the complete, top-of-the-line system for $2,972.

Downloaded Weather Information for Windows

Cyclogenesis
327 Rivershore Drive
Elk Rapids, MI 49629
(616) 264–6005
fax: (616) 264–6991

Cyclogenesis is a Windows-based weather information provider. By dialing a number your

CHAPTER 13: WEATHER FORECASTS

computer can download hundreds of different weather products including satellite images, difax charts, and text products. The images are in full color! These products are updated many times a day, so they're as current as you can get.

Chapter 14
Books about the Weather

Many books about the weather have preceded *The Weather Sourcebook*. Yet their focus, content, and coverage are so different from this book that you likely will find many of them irresistible. In this chapter we're going to tell you about some of the best weather books available. We've listed books for young readers separately, grouping them together in their own section.

You can find books on just about any aspect of the weather that interests you. Some are collections of interesting—or frightening—weather stories. You can learn about the history of weather and about weather wisdom from some. Other books present technical weather facts in simple lessons using plain language and basic experiments that enhance your learning. You can find chronicles of top weather news stories dating back to the 1850s.

Of course, in the visually oriented world of the 1990s, you can expect to learn a lot about the weather from videos as well as books. Chapter 15, "How to Learn More about the Weather," includes sources for interesting and educational videos on the weather.

Here's a listing and brief summary of other books about weather that we've found worthy, listed alphabetically by the author's last name, in three groups: twenty-eight books for adults or teenagers, seventeen books for young readers, then a special section at the end of the chapter that lists a few highly attractive coffee-table editions that would make excellent gifts for weather aficionados.

Adult-Reader Books about the Weather

Atkinson, B. W. and Alan Gadd. *Weather* (Weidenfeld & Nicolson, 1987)

Featuring a foreword by celebrity meteorologist Dr. Frank Field, this illustrated book is a collector's item that would be an excellent gift selection. It presents the latest expert knowledge available, in two sections. Section 1 explains how weather is created and distributed around the planet. Section 2 details the revolution going on in the science of forecasting: How supercomputers process incredible amounts of data from weather satellites, radar networks, specially designed aircraft, upper-atmosphere balloons, and the whole infrastructure of monitoring and forecasting techniques that can produce twenty-four-hour global forecasts at fifteen levels in the atmosphere in about four minutes!

Bohren, Craig F. *Clouds in a Glass of Beer* (John Wiley & Sons, 1986)

Bohren presents a collection of simple experiments that demonstrate atmospheric physics by letting you observe and reproduce natural phenomena with simple materials at home. Informative and engaging, it brings the subject of weather down to earth with offbeat, everyday examples and easy-to-follow experiments. Both professionals and amateurs can learn from these lessons. Not for young children, these experiments should be performed by or with the supervision of someone knowledgeable in the field of meteorology.

———. *What Light Through Yonder Window Breaks* (John Wiley & Sons, 1991)

This book is a sequel to Bohren's first book on atmospheric-physics experiments. Here he guides readers through the curious world of phenomena encountered each day. Written for an audience without specialized knowledge or expensive equipment, it investigates numerous questions, such as, Why does dew form on the *inside* of a window? What causes highway mirages? Bohren answers these and other questions using illustrations, photographs, and easy-to-perform experiments. This book will get you into the habit of seeing everyday things in a different light.

Caplovich, Judd. *Blizzard! The Great Storm of '88* (VeRo Publishing Company, 1987)

Caplovich documents the famous blizzard that inundated the eastern United States in March 1888. It's a detailed and fascinating account of one of the worst storms in the history of American weather, presented in a coffee-table–size format.

Carpenter, Clive. *The Changing World of Weather* (Facts on File, 1991)

In this illustrated book Carpenter provides a lively, readable account of the weather, showing that the Earth and its atmosphere are forever changing and always fascinating. Produced with the expensive look and feel of a coffee-table edition, it is an especially timely explanation of why the world has seen so many weather records in the last few years. It addresses the question, "What truly is happening to our global weather patterns?" You'll find terrific discussions of the ozone layer, the greenhouse effect, desertification, the destruction of our tropical rain forests, population pressures on the land, and alternatives to burning fossil fuel. This is an especially nice gift volume.

Day, John A. and Vincent Schaefer. *Peterson's Field Guides: Clouds and Weather* (Houghton Mifflin Company, 1991)

Dr. Schaefer is more than an expert on clouds. He's an expert on clouds who has done much more than observe them—he has also helped to make them. In 1946 Dr. Schaefer made the highly significant contribution of discovering the principle used today to seed clouds. In addition to thoroughly explaining cloud seeding, this book is loaded with color photographs taken of cloud formations all over the world. It also includes a detailed explanation of the full precipitation cycle in nature. This is an easily transportable, handy pocket guide that all weather lovers will want in their collections.

Erickson, Jon. *Violent Storms* (Tab Books, 1988)

A natural addition to the library of any reader interested in the weather. Erickson chronicles violent-weather phenomena, including their development, their history, and how past storms have affected people's lives.

Fitzgerald, Ken. *Weathervanes and Whirligigs* (Clarkson N. Potter, Inc., 1967)

A painstaking effort by the author, this book features more than 180 of his own illustrations. Fitzgerald truly begins at the beginning, with a discussion of the earliest weather vane for which any description exists: the Triton vane on the Tower of

CHAPTER 14: BOOKS ABOUT THE WEATHER

The Winds in Athens, dating from 48 B.C. He methodically tracks their history through time, up to today's modern vanes, which are so highly developed that they can detect anything other than perfectly still air. With his inclusion of wind toys, Fitzgerald will introduce you to the pleasure and delight of a good-humored native art.

Forrester, Frank H. *1001 Questions Answered About the Weather* (Dover Publications, Inc., 1981)

This title first appeared in 1957 but was completely revised in 1981. The new edition is well illustrated and is an excellent primer for amateur weather watchers, yet it can also serve as a valuable reference for professional meteorologists. It is structured into a reader-friendly question-and-answer format that makes it a breeze to locate topics that interest you. Forrester presents both factual and fun information about the weather, with more than 125 illustrations. For example, on pure scientific facts, he covers how storms originate, various weather instruments, and weather patterns. You can learn about how storm names originate and how weather instruments were invented and how they have evolved. It even can help you learn about careers in the field of meteorology.

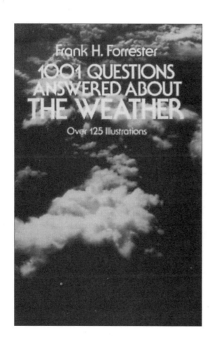

Hardy, Ralph, Peter Wright, John Kingston, and John Gribben. *The Weather Book* (Little, Brown, & Company, 1982)

A rich collection of historic material complements this scientific reference book. It covers basic information about the technical details of weather, such as how weather systems originate; how storms develop, travel, and dissipate; and how weather forecasters perform their craft. The book is embellished with color photographs, illustrations, and historical monographs.

Heuer, Kenneth. *Rainbows, Halos, and Other Wonders: Light and Color in the Atmosphere* (Dodd, Mead & Company, 1978)

This is an amazing collection of the most awe-inspiring spectacles of nature. Some, like the Brocken specter and the famed fata morgana, are so unearthly that they seem to belong to the world of the supernatural. Yet the author offers simple explanations, photographs, and drawings that illustrate the strange tricks that comprise this field, called *meteorological optics*. Some are commonly witnessed and easily explained, such as rainbows. Others, like the fleeting green flash at sunset, last but seconds and are seen by few. Some, like the rarely seen Cellini's halo, can be viewed regularly once you know how and under what conditions they exist. If you've ever wondered why the stars twinkle or if you can walk under the arch of a rainbow, this book will be rewarding reading.

Hughes, Patrick. *American Weather Stories* (U.S. Department of Commerce, 1976)

Here's a treasure trove of little-known anecdotes taken from the pages of American weather history. For example, the old adage about every cloud having a silver lining rings true in the story about how the otherwise devastating New York City blizzard of 1888 brought much-needed work to the city's unemployed people. Also learn about the little known fact that in 1492, when Columbus sailed the ocean blue, he encountered a hurricane! This book brings weather history to life.

———. *A Century of Weather Service* (Gordon & Breach Science Publishers, 1970)

Patrick Hughes chronicles the history of the National Weather Service in a detailed, comprehensive record.

Kotsch, Admiral William J. *Weather for the Mariner,* Third Edition (United States Naval Institute Press, 1983)

Rear Admiral Kotsch's classic mariner's guide to the weather contains nearly all you'll ever need to know about the advice, warnings, and safety tips that modern science can provide. Whether you are a weekend sailor, an oceangoing yachtsman, or a professional mariner, this book will provide a wealth of useful information. It is designed to help you recognize signs and conditions that can lead to maritime disasters. The knowledge you can gain from it will give you the confidence to enjoy safe boating.

Lee, Albert. *Weather Wisdom* (Doubleday, 1976)

Lee's book includes many historical anecdotes but basically is a manual on observational weather forecasting. He'll hold your interest with a weather version of "Ripley's Believe It or Not," in which Lee presents old weather yarns and sayings and adds his explanation or opinion on their veracity.

Lee, Sally. *Predicting Violent Storms* (Franklin Watts, 1989)

With this book you can learn the latest techniques for forecasting and predicting violent storms, including hurricanes, tornadoes, thunderstorms, and blizzards. The book begins with easy-to-follow explanations of the causes of violent storms. Next you'll learn about tools used in weather forecasting, such as barometers, hygrometers, weather maps, radar, and advanced satellites. Also described are the workings of the National Weather Service and how it processes more than 50,000 weather observations each day from land, sea, and sky.

Lockhart, Gary. *The Weather Companion* (John Wiley & Sons, 1988)

Here's a complete collection of meteorological history, science, legend, and folklore in one volume. Extensively illustrated with black-and-white line drawings, it features a detailed bibliography and is indexed. Among its many topics are bizarre stories from weather past, amazing weather tools, interesting weather phenomena, and unusual weather predictors. This book should appeal to naturalists, sportsmen, gardeners, and sky watchers of all ages—and anyone who ever has had an arthritic knee before a big rain.

Ludlum, David M. *The American Weather Book* (Houghton Mifflin Company, 1982)

This is one of the most complete, detailed treatises on American weather information available anywhere. Enhanced by a rich array of charts, graphs, and photos, this book presents region-by-region descriptions of all kinds of weather phenomena and records. For easy reference its chapter layout walks you through a calendar year, month by month, each chapter dedicated to the predominant weather patterns of the season it represents. Each month is introduced with a summary of its weather, followed by a complete, day-by-day listing of records and interesting weather occurrences.

———. *The Audubon Society Field Guide to North American Weather* (Alfred A. Knopf, Inc., 1991)

This is the ultimate weather reference book. It's a stunning collection of more than 200 pages that feature nearly 400 color plates. Each color photograph includes a caption with a page-number reference to explain the phenomena in the photo. You can match the sky to a color plate in the book, look at the reference page, and learn all about the cloud formation, including its significance in predicting weather. If you buy but one weather book, let this be the one.

———. *The Weather Factor* (Houghton Mifflin, 1984)

This offering from Ludlum compiles a collection of little-known facts that show how the weather has affected American history from colonial to modern times. You'll be amazed at how weather works to shape our world in surprising ways: in politics, by altering the outcomes of elections and wars; in sports, by changing fortunes; in the transportation industry, by determining the timing and the location of many firsts in aviation; and much more.

Lydolph, Paul E. *The Climate of the Earth* (Rowman & Allanheld, Publishers, 1985)

This is an exhaustive, detailed reference volume that contains a complete description of every climate on Earth. An extensive index enhances its reference value. It includes suggested "further readings" on a wide variety of topics related to climate: circulation systems, microclimates, meso-scale systems, climate and vegetation, agroclimatology, applied climatology, and urban climate.

Ramsey, Dan. *How to Forecast Weather* (Tab Books, 1983)

This is a simple, illustrated, hardcover guide about learning how to forecast weather accurately. An entertaining book on how you can be an amateur meteorologist for fun and profit, this book goes beyond the expected discourse on how the weather works and how it can be predicted with standard instruments. It presents the human side of how people are affected by weather and how human reactions to changes in conditions can predict weather.

Roan, Sharon L. *Ozone Crisis: The 15-Year Evolution of a Sudden Global Emergency* (John Wiley & Sons, 1990)

Anyone interested in understanding the environmental policy issues of the 1990s will find that Roan has created a well-researched, well-balanced, and informative book written in an easy-to-read, journalistic style. Taken straight from the headlines of today's news, here is the full story of the discovery of an ecological catastrophe that is destroying the shield of ozone gas that protects us from the sun's harmful ultraviolet rays. Beginning with the discovery in 1973 of the now-famous "hole in the ozone layer," Roan chronicles the corporate lobbying and government cover-ups that obstructed the early warnings of the discoverers. She concludes with a study of the actions taken to date by world governments to reduce the effects of the ozone crisis.

Rubin, Louis D. and Jim Duncan. *The Weather Wizard's Cloud Book* (Algonquin Books of Chapel Hill, 1984)

In the early 1960s a book entitled *Forecasting the Weather* was a pioneer in lay weather forecasting. This book is a reader-friendly revision written in simple language that makes weather easily understandable as well as interesting. You won't need a scientific background, either, as the authors clearly explain key weather terms. This is an excellent source book for amateur weather watchers, with its wonderful photographs of clouds and simple definitions of the different types shown. It's subtitled, "How You Can Forecast the Weather Accurately and Easily by Reading the Clouds."

Sanders, Ti. *Weather Is Front Page News* (Icarus Press, 1984)

"Extra! Extra! Read all about it! The Greatest American Weather Disasters!" Here are detailed stories behind the best of the best in front-page weather headlines about the greatest weather disasters in American history. Sanders presents stories going back to 1850 that relate the human side of cold waves and heat waves, floods and droughts, and much, much more.

———. *Weather: A User's Guide to the Atmosphere* (Icarus Press, 1985)

Sanders presents a completely different sort of

book from the previous listing. This one focuses on how the weather works in a reader-friendly, wide-ranging overview of the science of weather phenomena.

Schaefer, Vincent and John Day. *Peterson's Field Guides: Atmosphere: Clouds, Rain, Snow, Storms* (Houghton Mifflin, 1981)

This book is filled with an incredible collection of photographs of every imaginable cloud type and weather phenomenon. Each photo is accompanied with reference text that describes its features, its significance in predicting and understanding weather, and how it is formed. Every weather watcher will want this book as a complete reference guide. Its small size, heavy-duty vinyl cover, high-quality paper, and rugged binding make it durable enough to withstand the rigors of repeated use in the field.

Young-Reader Books about the Weather

Ardley, Neil. *The Science Book of Weather* (Gulliver Books, Harcourt Brace Jovanovich, 1992)

Here is a modern, up-to-date weather book for young readers. It begins with an easy-reading explanation of what weather is and how it affects humans, animals, and plants. Then, using brilliant colorful photographs, Ardley dedicates two or three pages to every type of basic weather phenomenon and includes one or two simple experiments that illustrate the principle in the text. Next to the photographs of these simple, home-built weather instruments, you also get photographs of corresponding equipment used by professional meteorologists.

Armbruster, Ann and Elizabeth A. Taylor. *Tornadoes* (Franklin Watts, 1989)

This book not only explains tornadoes but also presents the information with the aim of quelling fears. It gives readers tips on how to recognize tornado conditions so that they can prepare and protect themselves in case one comes. The authors also include some science projects and related activities that illustrate how tornadoes are formed. In the last chapter they list other references on tornadoes for readers who may be writing reports on them for a science class.

Bramwell, Martyn. *Earth Science Library: Weather* (Franklin Watts, 1987)

Earth Science Library is a full-color series that explores the world of physical geography. Each book explains the forces that shape the Earth and describes its landscapes and features. This *Weather* volume features outstanding color photographs supported by explanatory diagrams. It could serve as an excellent reference source for a junior high or high school student doing a report on weather. In addition to colorful photos and reader-friendly text that explains how weather works and can be forecast, it includes a discussion of how humans are affecting changes in climate and the weather.

Branely, Franklyn M. *Flash, Crash, Rumble and Roll* (Thomas Y. Crowell, 1985)

Do you know a child who is afraid of thunder or lightning? This book is perfect for even the youngest or most fearful child. It features illustrations by Barbara and Ed Emberley that embellish the text so that it not only explains how weather works but why it includes such scary sights and sounds. It takes the fear out of the things that won't hurt you and illustrates how to protect yourself from the weather that can be dangerous.

DeWitt, Lynda. *What Will the Weather Be?* (Harper Collins Publishers, 1991)

Here's a treasure trove for children from the "Let's-Read-and-Find-Out" science book series. It is the perfect gift book. Every page is illustrated with a color picture by Carolyn Croll. The result is a simple yet highly informative introduction to weather forecasting that children will find interesting. It not only helps them understand how weather forecasts are created but also how we can use them and why doing so is important.

CHAPTER 14: BOOKS ABOUT THE WEATHER

Flint, David. *Weather and Climate* (Gloucester Press, 1987)

This book would make a memorable gift for any child, presenting young-reader–friendly explanations of both weather and climate. Every page is beautifully illustrated with color charts, drawings, or photographs. It covers trends in weather, from those that affect us from day to day to the various climates that influence whole areas of the world. It is richly sprinkled with "hands-on" sidebars featuring projects to enhance learning and "did you know?" sidebars with special information for added interest.

Gibbons, Gail. *Weather Forecasting* (Macmillan Publishing Company, 1987)

Gibbons presents a behind-the-scenes look at a modern weather station to show how professional weather forecasters at the National Weather Service practice their craft. With clear, concise text and crisp, colorful pictures, she describes the fascinating equipment used to track and gauge the constantly changing weather.

———. *Weather Words and What They Mean* (Holiday House, 1990)

Everyone talks about the weather, but young readers often do not understand all the words used. Gibbons explains where fog, clouds, frost, thunderstorms, snow, fronts, and other weather-related phenomena come from. Common terms that are not commonly understood are clearly defined. Each is illustrated with colorful pictures that graphically show what the term represents. A list of curious weather facts is also provided.

Kahl, Jonathan D. *Weatherwise: Learning About the Weather* (Lerner Publications Company, 1992)

Nearly every page of this book has a color chart, drawing, or photograph to help young readers understand the text. Most words that may be unfamiliar are printed in boldface type and listed in a glossary in the back of the book. A complete index of topics also is provided. The author begins by explaining the weather cycle from a global perspective that includes simple discussions about the greenhouse effect and why the Earth's climates differ. In the book's closing section, young readers can learn how to read a weather map and how professionals forecast weather.

Kramer, Stephen. *Lightning* (Carolrhoda Books, Inc., 1992)

Until you see this book, you may never have imagined that lightning could be so complex. Yet this author, with the help of a lavish array of incredible photographs by weather photographer Warren Faidley, makes it interesting and simple. Detailed color charts enhance Kramer's concise explanations. The photographs include rare computer-enhanced shots that show the step-by-step development of a full lightning stroke.

Lampton, Christopher. *Drought* (The Millbrook Press, 1992)

Lampton investigates the causes and the disastrous effects of drought, giving the history of some of the severest droughts on record in the United States and elsewhere. Beginning with emotional photographs of "Okies" on their heart-wrenching Westward trek, and continuing through the latest developments in the Sahel area of Africa, this book gives a face and a heart to the effects of drought. Using simple charts and photographs to accompany reader-friendly text, it ends with an excellent section on the technical aspects of drought and its causes.

Mandell, Muriel. *Simple Weather Experiments With Everyday Materials* (Sterling Publishing Company, 1990)

This book is not just for kids. Here you can find truly simple experiments that graphically explain weather phenomena even as complex as the Coriolis effect. This volume is so up to date it even shows how to simulate the creation of acid rain. The book is sprinkled throughout with listings of interesting weather facts and records. Different topics are organized into four separate sections to make reference easy: "Warming Up" covers temperature facts and

experiments; everything about air and wind is grouped into "Whirling Winds and Gentle Breezes"; experiments about precipitation are found in "Water, Water, Everywhere"; and the last section shows how to build and use a rather complete weather station employing common household items and simple techniques.

McVey, Vicki. *The Sierra Club Book of Weatherwisdom* (Sierra Club Books/Little, Brown & Company, 1991)

The author uses a winning combination of dramatic stories, fascinating information, and imaginative hands-on activities, games, and experiments to make a potentially complex subject easy and fun to learn. Using young fictional characters from different parts of the world whose lives are keenly affected by the weather, she shows readers how they, too, can become weatherwise by learning basic weather principles and paying close attention to the world around them. She offers vivid explanations of how the Earth's rotation and orbit are linked to the seasons and climates, what makes wind and rain, how storms develop, and much more. It is nicely illustrated with pencil drawings by Martha Weston.

Pettigrew, Mark. *Weather: Science Today* (Gloucester Press, 1987)

Pettigrew has written a versatile reference book for older children. He provides color pictures and charts on every page along with easily understood descriptions of the weather. He includes many simple experiments, including instructions on creating an environmentally friendly home weather station that can be built using recycled, common household materials.

Simon, Seymour. *Storms* (Morrow Junior Books, 1989)

This author very simply describes the atmospheric conditions that create major storms. The book covers thunderstorms, lightning, hailstorms, tornadoes, and hurricanes. Nearly every page is illustrated with color photographs from the National Center for Atmospheric Research. In addition to showing how these storms are created, the book explains how violent weather affects the environment and people.

Sugarman, Joan. *Snowflakes* (Little, Brown & Company, 1985)

The fleeting beauty of the microscopic world of snowflakes is captured in this informative, illustrated volume. The text describes in reader-friendly language how the seemingly infinite varieties of snow crystals are formed. Detailed pen-and-ink drawings by Jennifer Dewey bring the shapes of stars, needles, feathers, and pyramids alive. In one chapter Sugarman tells you how to capture and preserve snowflakes so you can enjoy their beauty up close.

Tannenbaum, Beulah and Harold E. *Making and Using Your Own Weather Station* (Franklin Watts, 1989)

This is the most complete collection of detailed instructions on building your own equipment for a home weather station. It describes various weather phenomena and includes a materials list and directions on building a weather unit to monitor each. Plans for a barometer, thermometer, sling psychrometer, rain gauge, snow gauge, wind vane, and anemometer are included. By showing how to use their sample recording sheet, the authors teach readers how to use the instruments to forecast weather accurately.

Coffee-Table Books about the Weather

Bishop, Robert and Patricia Coblentz. *A Gallery of American Weathervanes and Whirligigs* (Bonanza Book, 1984)

Here is a beautiful volume that celebrates weather vanes and assorted wind devices. Consider this book as a great gift for anyone interested in American arts and culture as well as weather buffs. It's divided into chapters that cover related devices, such as weather vanes with horses, angels, farm animals, statues, or nautical themes. You'll be amazed at the stunning array of American weather vanes and at the imaginations of the people who have created them. This book consists mostly of color photographs, but a few pages of text at the beginning introduce the collection.

Cosgrove, Brian. *Eyewitness Books: Weather* (Alfred A. Knopf, 1991)

Hundreds of stunning photographs and lively captions present an entirely fresh look at the wonders and workings of the world around us. It's like a mini-museum between the covers of a book. Every page is a treasure trove of multiple-color photos, drawings, and charts. Some of the photographs show beautifully crafted models created especially for this book to illustrate complex weather phenomena. A unique feature makes cloud patterns easy to understand: throughout the book a series of large color photographs of cloud patterns are overlaid with drawings of mini-weather instruments that show the barometric pressure, cloud cover, and wind conditions that created the patterns in the photos. Kept on display in your home or in an office waiting room, this book can provide years of enjoyment for weather fans of all ages.

Lambert, David and Ralph Hardy. *Weather And Its Work: The World of Science* (Macdonald & Company, Ltd., 1987)

Every page of this highly interesting book has color drawings, charts, or photographs to embellish the concise, well-written explanation of nearly every known weather phenomenon. It is an excellent and enjoyable reference for learning about the weather. Terms that might not be widely understood are printed in italics to denote their inclusion in the glossary. Rather than giving a lot of details about weather, the authors depict weather's impact on our planet's geography and on its plant and animal life.

Chapter 15
How to Learn More about the Weather

Few things in life are more satisfying than studying your most passionate interests. If you've been smitten by the weather bug, you'll certainly want to increase your weather knowledge. Fortunately, the weather is a topic in which you can become completely immersed. The past decade has brought an explosion in the store of weather information that you can tap.

As we have become more aware of how weather works and how it affects our daily lives, interest in learning about weather has exploded. Weather events and weather data are perfectly suited to being captured by recent high-tech developments in computers and video recorders. It now is easy to keep track of weather data and forecasts. And there are more good reasons than ever to track this information because, as our understanding of weather phenomena has increased, we are able to use the data we gather more effectively to alter our plans and behavior patterns as dictated by the weather.

How does this weather bug bite, and what do those smitten by its allure do once called? Here's a look at the inspiration behind one weather enthusiast and what he's done with his excitement about the study of weather.

"Many Paths Can Be Followed"

Many paths can be followed along the road to understanding and interpreting the weather. Some become professional meteorologists. Others may get involved in weather photography. Still others may chase tornadoes during the spring thunderstorm season.

For American Association of Weather Observers (AAWO) member Keith Haley, weather interest seems to have been the result of a variety of factors.

"It was a combination of things. My mother, father, and Junior High school science teachers all were a source of encouragement. My father—who sailed the Great Lakes during the mid and late 1940's—fueled my interest by explaining to me how storms would come up very fast on Lake Michigan. In addition, he explained how Indians could predict the weather for the next season by looking up at the northern lights."

Not unlike many of you, Keith also had his interest rekindled after witnessing his first funnel cloud as it passed over his home in May of 1971.

"I noticed how things got very quiet. The air was nearly still and my animals were very upset. There was a lot of electrical interference on the radio, and the sky looked very disturbed and dark. I noticed that in the southwest quadrant of the sky, a large, dark, moving object was coming my way. At first I had no idea it was a tornado!"

While Keith is no stranger to the study of weather, he is also fascinated by electronics and computers, a path he followed through his degree from Western Michigan University in Electricity and Electronics Industrial Education in 1977.

For the past few years, the Belleview, Michigan, resident has spent a great deal of time writing and updating software for maintaining and calculating weather records, and for use as an aid to predicting the weather, a job he finds very rewarding.

Keith is but one of a number of AAWO members who have found that a combination of computers and weather is a meaningful path to follow. Another you might be familiar with is Minnesota's Dan Lipinski. There are still others who also see this as a real need for the weather community and continue working on our behalf.

Keith has used his knowledge and skills to help others learn about the world of weather and computers. He is a local Skywarn spotter and maintains his own amateur radio license, N8OCC.

But one thing he really enjoys, is sharing his interests with students at local schools and organizations.

"I talk about what is involved in the making of a weather forecast and share with young people how the weather affects history, ecology, and the environment."

And yes, he shows them how his software can help others predict and record the weather.

Keith Haley was named our At-Large Observer of the Year in 1990. It's easy to see why he was selected.

His path has helped others learn more about our fascination with the atmosphere. He has no doubt inspired some young students to learn more about it too.

—By Debi Iacovelli, Keith Haley, and Steve Steinke.
Reprinted with permission from the *American Weather Observer*.

Perhaps Keith Haley's story and his work will inspire you, too, to jump into the intriguing and ever-changing field of weather. If so, then the rest of this chapter is for you as we present:

- institutions that offer weather courses
- videos that give you close-up looks at some spectacular weather phenomena and the technology that goes into tracking and observing them
- publications that will keep you up to date with the latest developments as our knowledge of weather continues to grow at exponential rates
- a brief guide to the world's only weather museum

The Weather Museum

The Museum of American Weather opened in Haverhill, New Hampshire, on June 7, 1992, on the grounds of the home of Roger K. Brickner. The museum manager, Matthew Madan, of East Topsham, Vermont, oversees the facility. It's the first and only privately owned weather museum in the world.

Only in its infant stages, the new weather museum continues to collect items for display. Yet it opened with a good initial array of weather items. Here's a sample of some of the exhibits the museum features in its displays:

- murals (a triptych) of the 1938 hurricane with weather maps and data
- murals of the Blizzard of 1888 with weather maps and data
- exhibit of Hurricane Hazel (1954) with data and weather maps
- weather photography (anyone with weather photographs, please feel free to contribute)
- a depiction of the various cloud types as mobiles hanging from the rafters of the museum
- stained-glass pendants that illustrate great weather events of the past
- a display of early weather records and engraved reproductions of conditions on top of Mount Washington

The museum also houses a library that stocks a wide variety of meteorological books, records, and videotapes of weather. The library/theater walls have a unique weather wallpaper—be sure to look for it. Dispersed here and there are old weather instruments. The library still is expanding, so if you have any of the items they carry that you would care to contribute to a worthy cause, please call Mr. Madan at (802) 439-5601.

This museum, dedicated solely to the weather, is for you. Weather enthusiasts always will be welcome, as will be your suggestions and your contributions. Only with your help can your museum prosper.

Weather Educational Institutions

Here's a listing of educational institutions, in both the United States and Canada, that offer curricula in the atmospheric, oceanic, hydrologic, and related sciences. You can write to any that interest you for complete information and brochures.

U.S. Universities and Colleges

Alabama
University of Alabama, Huntsville

Alaska
University of Alaska, Fairbanks

Arizona
University of Arizona, Tucson

California
Naval Postgraduate School, Monterey
San Francisco State University, San Francisco
San Jose State University, San Jose
Scripps Institution of Oceanography, La Jolla
University of California, Davis
University of California, Irvine
University of California, Los Angeles
University of Southern California, Los Angeles
Stanford University, Palo Alto

Colorado
Colorado State University, Fort Collins
Metropolitan State College, Denver
University of Colorado, Boulder
University of Denver, Denver

Connecticut
Yale University, New Haven

Delaware
University of Delaware, Lewes

Florida
Florida State University, Tallahassee
Nova University, Dania
University of Florida, Gainesville
University of Miami, Miami

Georgia
Georgia Institute of Technology, Atlanta

Hawaii
University of Hawaii, Honolulu

Illinois
Northern Illinois University, de Kalb
University of Chicago, Chicago
University of Illinois, Urbana-Champaign
Western Illinois University, Macomb

Indiana
Indiana University, Bloomington
Purdue University, West Lafayette

Iowa
Iowa State University, Ames

Kansas
University of Kansas, Lawrence

Louisiana
Northeast Louisiana University, Monroe

Maryland
Johns Hopkins University, Baltimore
U.S. Naval Academy, Annapolis
University of Maryland, College Park

Massachusetts
Harvard School of Public Health, Boston
Harvard University, Cambridge
Massachusetts Institute of Technology, Cambridge
University of Massachusetts, Lowell
Woods Hole Oceanographic Institution, Woods Hole

Michigan
Central Michigan University, Mt. Pleasant
University of Michigan, Ann Arbor

Minnesota
St. Cloud State University, St. Cloud

Mississippi
Jackson State University, Jackson
Mississippi State University, Mississippi State

Missouri
University of Missouri, Columbia
University of Missouri, Rolla
Saint Louis University, St. Louis

Nebraska
Creighton University, Omaha
University of Nebraska, Lincoln

Nevada
University of Nevada–Desert Research Institute, Reno

New Hampshire
Plymouth State College, Plymouth
University of New Hampshire, Durham

New Jersey
Kean College of New Jersey, Union
Princeton University, Princeton
Rutgers University, New Brunswick

New Mexico
New Mexico Institute of Mining and Technology, Socorro

New York
City College of New York, New York
Columbia University, New York
Cornell University, Ithaca
New York University, New York City
State University of New York, Albany
State University of New York, Brockport
State University of New York, Oneonta
State University of New York, Oswego
State University of New York–Maritime College, Bronx

North Carolina
North Carolina State University, Raleigh
University of North Carolina, Asheville

North Dakota
University of North Dakota, Grand Forks

Ohio
Ohio University, Athens
Ohio State University, Columbus

Oklahoma
University of Oklahoma, Norman

Oregon
Oregon State University, Corvallis

CHAPTER 15: HOW TO LEARN MORE

Pennsylvania
Drexel University, Philadelphia
Millersville University of Pennsylvania, Millersville
Pennsylvania State University, University Park

Puerto Rico
University of Puerto Rico, Mayagua

Rhode Island
University of Rhode Island, Narragansett

South Dakota
South Dakota School of Mines and Technology, Rapid City

Texas
Rice University, Houston
Texas A&M University, College Station
Texas Tech University, Lubbock
University of Texas, Austin

Utah
University of Utah, Salt Lake City
Utah State University, Logan

Vermont
Lyndon State College, Lyndonville

Virginia
Old Dominion University, Norfolk
University of Virginia, Charlottesville

Washington
University of Washington, Seattle
Washington State University, Pullman

Wisconsin
University of Wisconsin, Madison
University of Wisconsin, Milwaukee

Wyoming
University of Wyoming, Laramie

Canadian Universities
University of Alberta, Edmonton
University of British Columbia, Vancouver
Dalhousie University, Halifax, Nova Scotia
University of Guelph, Ontario
McGill University, Montreal
McMaster University, Hamilton, Ontario
University of Saskatchewan, Saskatoon
University of Toronto, Ontario
York University, North York, Ontario

National Weather Service

8060 13th Street
Gramax Building
Silver Spring, MD 20910
(301) 427–7622

The ultimate source of weather information in the United States is the National Weather Service. All weather roads lead here. The NWS gathers daily weather information from more than 10,000 sources: weather buoys, ships, land-based observation stations, radar sites, weather satellites, international observers, and volunteer domestic observers. This enormous databank of weather information is processed to create forecasts that are distributed worldwide.

The NWS offers a comprehensive catalog of free brochures that cover nearly every known weather topic, including thunderstorms, hurricanes, and tornadoes. These informative brochures tell you how these storms form and how to predict and track them, and they provide safety tips on how to respond when you are alerted that one is approaching. The NWS also makes its databanks available to the public for electronic downloading (this information, however, is not free, and the fees are high enough that few amateurs will be interested).

Weather Videos

After the Warming **(produced by Maryland Public Television, 1990)**
120 minutes (in two parts), $185 plus $6.00 shipping and handling
Ambrose Video Publishing
1290 Avenue of the Americas
New York, NY 10104
(800) 526-4663

Maryland Public Television teamed with Film Australia and Wisemans and Electric Image (U.K.) to produce this PBS special on the greenhouse effect. The story takes the perspective of someone living in the year 2050 and features the wry humor of host James Burke.

Part I, entitled "The Fatal Flower," shows the relationship of climatic changes to the development of civilization, dating back to the last ice age, which forced humans to learn hunting and enabled them to migrate to America and Australia. After the glaciers retreated when the ice age ended, humans settled down a bit more and turned to farming. The story continues on to track human travel and lifestyles through Roman times, the Viking era, the "little ice age" (in the 1300s), and the Industrial Revolution.

Part II, "The Secret of the Deep," takes a forward view, predicting an apocalyptic future as global warming melts the ice caps, producing coastal flooding, as droughts produce widespread famine, and as tremendous tropical cyclones cause massive loss of life. Fortunately, Part II takes a turn for the better and portrays a future in which humankind takes positive action that brings the greenhouse effect under control.

There's a contradiction between the two parts, however. Part I thoroughly explains the dramatic changes in the Earth's climate before humankind had any input; yet Part II puts the entire burden for expected future changes on adverse human affects on the environment, with no allowance made for such natural changes as have occurred over the ages.

Chasing Down a Dream **(produced by Tim Dorr and Jack Corso, 1991)**
120 minutes, $34.95
Jack Corso
P. O. Box 650
Harrison, NY 10528

The highlight of this noticeably amateur production is footage of an F4 tornado that Tim Dorr and Jack Corso caught on tape in its initial stages and followed through until nightfall. The funnel stayed on the ground for half an hour. Additionally, there are some good shots of supercell storms and their accompanying features. There also is good footage of other Plains storms, thunderstorms in Florida and New York, and Hurricane Bob as it struck Rhode Island.

The soundtrack is this tape's major shortcoming, featuring a car-radio background and numerous expletives, blurted out in the excitement of the chase. Also, the editing is not tight, resulting in repetitious scenes and footage of empty, black night sky. This tape works overtime to stretch to two full hours, but Dorr and Corso remain enthusiastic throughout.

Chasing the Wind **(produced by Martin Lisius, 1991)**
27 minutes, $20
Plano Television Network
Attn: Martin Lisiur
P. O. Box 860358
Plano, TX 75086-0358
(214) 578-7179

CHAPTER 15: HOW TO LEARN MORE

A sampling of the dedicated work of serious storm chasers on the southern Great Plains are captured in this video, which deals with both professionals and amateurs. You'll not only get the excitement of seeing a successful chase unfold but you'll also get an insider's taste of real life for these people as you see their daylong efforts end under clear, wide-open skies with no storm in sight. This inside look lets you learn about the psyche of storm chasers and hear about the dangers they face in their close calls with nature's most unstoppable force.

Near the end you watch them collect on the dues they've paid as April 26, 1991, unfolds. That memorable day was a textbook example of the weather conditions that storm chasers seek, drawing them out by the hundreds from all across Oklahoma and Kansas. It ended with a supercell thunderstorm that produced a tornado in Wichita, Kansas, captured on this video.

Chase To Live (produced by Gene Rhoden, 1993)

120 minutes, $29.95
Chase to Live
707 Timberdell Road
Norman, OK 73072

This is a unique chase video. It differs from other storm videos because it is suitable for training spotters/chasers. It illustrates the hazards associated with chasing, such as tornadoes that become rain-wrapped, baseball-size hail striking the chase vehicle, a very close cloud-to-ground lightning strike, and core-punching a violent tornado. It shows other hazards that chasers/spotters might encounter, such as flooded roads, driving in heavy rain and hail, low-contrast tornadoes, multiple-vortex and merging tornadoes, and spectacular lightning displays. It also features the development and evolution of a supercell, from the towering cumulus stage through the tornado stage. The production quality is excellent. This tape is highly recommended, especially to acquire an insider's knowledge of tornadoes.

Dangerous Storms (produced by Warren Faidley, 1992)

60 minutes, $29.95 plus $3.95 shipping and handling
Texas Weather Devices
P.O. Box 309
Cresson, TX 76035

Here is a well-structured presentation, featuring four separate sections: Hurricanes, Tornadoes, Lightning, and the Spotter's Guide for severe-thunderstorm spotters. Each section is "bookended" with a brief description of the topic phenomenon at the opening and a lesson on storm-related safety rules at the end. Between the bookends producer Warren Faidley includes both short action videos and still shots of some of his award-winning photographs.

The Tornado section presents close-up video of the famous 1991 Kansas tornado tearing apart the Terradine Country Club near Wichita and a June 2, 1990, tornado in Illinois. The Hurricane section features shots of the 150-mph winds of Hurricane Hugo lashing Puerto Rico. Faidley features some of his famous stills in the Lightning section.

Faidley's intention with this video is to educate the public about severe storms; toward that end he offers the safety tips with each section and the Spotter's Guide. Examples of his excellent teaching tools are a 3D computer simulation of a developing severe thunderstorm and a good explanation of supercell thunderstorms. The Spotter's Guide includes footage of an NSSL–chase-team encounter with a tornado plus still photos of tornadoes, tornado-look-alike clouds, wall clouds, and thunderstorms.

Hurricane Andrew: Satellite and Radar Sequences (produced by the Cooperative Institute for Meteorology Satellite Studies, 1992)

12 minutes, $20
Chris Velden
Space Science and Engineering Center
University of Wisconsin–Madison
1225 West Dayton Street
Madison, WI 53706

This short clip provides excellent infrared imagery of Hurricane Andrew, but you get little more than the pictures and a few captions; there is no narration. Other shots include limited visible imagery, water-vapor imagery that reveals internal steering currents, and radar images of the storm's passage across the Gulf of Mexico. All these images give a clear picture of the factors that caused the storm to grow so rapidly to be classified as a Category 4 hurricane.

Hurricanes of the 1980s (produced by Richard Horodner, 1991)

82 minutes, $29.95
Richard Horodner
9961 S.W. 156 Terrace
Miami, FL 33157

Watch this video and you'll feel that you are inside a hurricane. Richard Horodner has compiled an excellent highlight video footage from the best of his 1980s hurricane chases. Fast-paced editing keeps the story moving and will hold your interest. You'll certainly get a feel for the dangers of being a storm chaser. This is one of Horodner's best.

Seven hurricanes are featured, beginning with Hugo in 1989, then Gilbert in 1988; Bonnie in 1986; Elena, Gloria, and Kate from 1985; and 1984's Diana. Each segment depicts the storm's track on a map and features satellite photographs and a review of the storm's effects. Hugo and Gilbert, being the largest of the storms covered, get the most coverage, with spectacular footage.

Planet Earth (produced by PBS, 1990)

Seven 60-minute tapes, $195.65 plus $5.00 shipping and handling
Easton Press Video
47 Richards Avenue
Norwalk, CT 06857
(800) 367-4534

You now can own the complete set of this Emmy Award-winning PBS series. Seven stunning videos show the quantum leap that our knowledge of the Earth's development has taken in the past decade. While not solely about the weather, the series clearly shows how our weather is a result of much larger events that have occurred over millions of years. In beautiful color photography, you'll see everything from views of hurricanes from the perspective of orbiting satellites to the depths of the ocean floor.

Through brilliant special effects you'll see continents move across the globe, parting and colliding like bumper cars. You also can plunge into the searing surface of the sun and ride the solar wind at 1 million mph.

You can learn how the Earth affects its own atmosphere, climate, and weather as volcanoes and its own rotation and magnetic field create atmospheric events. The series also shows how other heavenly bodies affect our planet's climate and weather as asteroids collide with the Earth's surface.

Spring Fever Live! (produced by Marty Feely, 1991)

86 minutes, $30, including shipping and handling
Marty Feely
5273 Halifax Drive
San Jose, CA 95130
(408) 379-1850

This tape is a documentary on storm chasing rather than a simple potpourri of severe-thunderstorm phenomena. Most of the chases begin with amateur storm chaser Marty Feely at the local NWS office, examining the day's synoptic situation and planning his chase. Then he takes us along for his exciting explorations.

This tape includes footage of tornadoes from three separate storms taken by Feely plus some good, distant shots of supercell storms. He also provides footage of a dust devil, large hail, urban flooding, rainbows, and nighttime shots of spectacular lightning displays.

Feely's amateur status shows up through a shortage of good narration, although he knows storms very well and points out various storm features as he pans the sky. He's good at positioning himself at the rear of the storm, where the real action occurs. The dearth of narration is offset somewhat by a written program that follows the tape.

Tornado Video Classics (produced by Tom Grazulis, 1992)

120 minutes, $48, including shipping and handling
Tornado Project
Box 302
St. Johnsbury, CT 05819
(802) 748–2505

This is not a chase video or a simple compilation of amateur footage with an amateur soundtrack. Here is a professionally produced educational video about tornadoes. Showing more than fifty tornadoes that struck twenty states from 1951 through 1992, this video gives a comprehensive understanding of these awesome storms in an organized historical perspective.

The video includes a detailed, forty-eight-page viewer's guide that features maps, diagrams, and experimental results. The video scenes are linked by number to the guide, so you can learn more than you otherwise would get from the excellent narration.

It includes, for the first time on video, the uncut documentary film *Approaching the Unapproachable*. This segment is an excellent teaching tool, recommended by *Science Teacher* magazine as "awe-inspiring." Another bonus is footage of dust devils, waterspouts, model tornadoes, and experiments plus tornado missile projectiles on rocket sleds. You also will get an inside look at the work of tornado experts Ted Jujita, from the University of Chicago, and the University of Oklahoma chase team leader Howard Bluestein, showing off their fabulous "storm toys" such as TOTO and a mobile Doppler radar station.

Weather Wise (produced by King Productions)

54 minutes, $29 plus $7.00 shipping and handling
King Schools
3840 Calle Fortunada
San Diego, CA 92123
(619) 541–2200

Learn how to make real-world use of weather information. This video will help you understand the dynamics of fog, thunderstorms, and frontal systems. It's produced by a highly acclaimed video aviation school, so its information is presented from the viewpoint of a pilot. Still, learning how pilots understand the weather will give the landlubber a new perspective on weather phenomena and how they affect everyone's life. This video will help you predict local weather conditions, showing you when to trust a forecast and when not to trust one. Pilots learn to be highly skeptical of all weather forecasts. The best gather their own data and make sure these are congruent with the forecasts they read—always a good policy, even if you're not flying.

Electronic Weather Data

World Weather Disc

WeatherDisc Associates, Inc.
4584 NE 89th Street
Seattle, WA 98115
(206) 524–4314

To use this incredible storehouse of information, you'll need a personal computer that has a CD-ROM drive. This disc is probably the most extensive single meteorological reference available. If you've got the equipment to handle it, you can get all this packed into a single compact disc:

- climatic data taken from thousands of weather observation stations around the world, with some data dating back to the 1700s
- average weather conditions at thousands of airports around the world
- daily weather data from hundreds of U.S. stations
- Climatography of the U.S. No. 20
- Local Climatological Data (LCD)
- U.S. Climatic Division Data
- U.S. monthly normals
- COADS ship data
- worldwide airfield summaries

These data sets encompass a large number of meteorological parameters, such as temperature, precipitation, heating/cooling degree days, freeze occurrence, drought and soil moisture, wind, sunshine, lightning, tornadoes, tropical storms, and thunderstorms. Weather professionals, scientists, educators, and amateurs alike can use and enjoy this complete weather knowledge base. Access software, with integrated graphics, is included in the $295 purchase price.

Weather Publications

The American Weather Observer

This monthly newspaper is for everyone who loves keeping up to date on the weather. Each issue has a host of regular features, such as:

- a cover story on the latest developments in the weather field
- a column by David Ludlum, reviewing America's weather from 50 and 100 years ago
- a column by the Reverend Robert Duane and Tom Johnston entitled "The Art and Fun of Weather Observing"
- updates on the most recent significant weather events
- articles on buying and using home weather equipment

There is nothing else like it on the weather market. Its link to the American Association of Weather Observers helps the paper represent the entire weather field. Therefore, it will keep you abreast of how the political winds are blowing in the world of the weather experts. The subscription rate for twelve issues is $21 and includes membership in the AAWO. Back issues cost 75 cents.

Weatherwise

This is one of America's popular weather magazine. It captures the power, beauty, and excitement of the ever-changing elements in vibrant color photographs and crisp, well-written articles. Popular annual features include the almanac issue, with insights from experts on the year's weather events. Several recent articles from *Weatherwise* have been reprinted in *The Weather Sourcebook*. *Weatherwise* is published bimonthly and costs $32 for individuals and $54 for institutions. Add $12 for postage outside the United States. Call (202) 362–6445.

Storm

America's newest weather magazine, *Storm*, edited by Dr. John Harlin of Northern Illinois University, premiered in 1992. Their goal is to present a "public" approach to weather topics. Articles from the premiere issue included "Monitoring Global Temperatures from Space," "Why We Need LAWS" (Laser Atmospheric Wind Sounder), "Television's Changing Climate," and "NWS 2000." Regularly featured departments include "Up Front," "Weather Wire," "Weather Science Notebook," a review of the previous month's storms, "Hot Products" (the latest weather-equipment reviews), "Desktop Weather," "Weather Pix," and the weather outlook for the coming month. Call (800) 547–0890 for subscription information ($24 per year) and newsstand availability.

NOAA Literature

U.S. Department of Commerce
National Oceanic and Atmospheric Administration
Rockville, MD 20852

The National Oceanic and Atmospheric Administration offers an amazing array of free or low-cost weather literature. It stocks material that describes and explains nearly everything you ever might want to know about nearly every weather phenomenon. Examples include how tornadoes form and how to protect yourself, how to track hurricanes and protect yourself against them, and how to participate in its *SkyWarn* program. The next time you wonder about how a weather event forms or how to protect yourself from it, contact the NOAA and request information.

CHAPTER 15: HOW TO LEARN MORE

Weather Learning Products

NASA Films and Filmstrips

The National Aviation and Space Administration maintains an extensive collection of films and filmstrips on the weather, covering all major violent-weather phenomena. Films are in 16-millimeter format and are lent at no cost. Filmstrips are for sale. Mark your envelopes "ATTN: NASA Teacher Resource Center" and mail to the address below that is nearest to you. Addresses are listed alphabetically by state.

Alabama Space and Rocket Center
Huntsville, AL 35807
(205) 837–3400

NASA Ames Research Center, Mail Stop 204-7
Moffett Field, CA 94035
(415) 694–6077

NASA Jet Propulsion Laboratory
JPL Educational Outreach, Mail Stop CS-530
Pasadena, CA 91109
(818) 354–6916

U.S. Space Foundation
P.O. Box 1838
Colorado Springs, CO 80901

National Air and Space Museum
Education Resource Center, P-700
Office of Education
Washington, D.C. 20560

NASA John F. Kennedy Space Center
Mail Stop ERL
Cape Canaveral, FL 32889
(305) 867–4090

Museum of Science and Industry
57th Street and Lakeshore Drive
Chicago, IL 60637

Parks College of Aeronautical Technology
Rte. 157 at Falling Springs Road
East St. Louis, IL 62201

University of Evansville
School of Education
1800 Lincoln Avenue
Evansville, IL 47714

Children's Museum
P.O. Box 3000
Indianapolis, IN 46206

Bossier Parish Community College
2719 Airline Drive
Bossier City, LA 71111

NASA Goddard Space Flight Center
Mail Stop 130-3
Greenbelt, MD 20771
(301) 344–8981

Northern Michigan University
Olson Library Media Center
Marquette, MI 49855

Central Michigan University
Mt. Pleasant, MI 48859

Oakland University
115 O'Dowd Hall
Rochester, MI 48063

Mankato State University
Curriculum and Instruction
Box 52
Mankato, MN 56001

Center for Information Media
St. Cloud State University
St. Cloud, MN 56301

NASA National Space Technology Laboratory
Building 1200
National Space Technology Laboratory, MS 39529
(601) 688–3338

University of North Carolina at Charlotte
J. Murrey Atkins Library
Charlotte, NC 28223

City College
NAC 5/208
New York, NY 10031

NASA Lewis Research Center
Mail Stop 8-1
Cleveland, OH 44135

NASA Industrial Applications Center
823 William Pitt Union
University of Pittsburgh
Pittsburgh, PA 15260

NASA Lyndon B. Johnson Space Center
Mail Stop AP-4
Houston, TX 77058
(713) 483–3455

NASA Langley Research Center
Mail Stop 146
Hampton, VA 23665
(804) 865–4468

Champlain College
174 South Willard Street
Burlington, VT 05402

University of Wisconsin at La Crosse
College of Education
La Crosse, WI 54601

Science, Economics, and Technology Center
818 West Wisconsin Center
Milwaukee, WI 53233

Weaterslam
Rencroc, Inc.
P. O. Box 588632
Seattle, WA 98188
(206) 242–2319

If you like to combine learning with fun, you'll love Weatherslam, an exciting, easy-to-learn, bidding card game where "the sky's the limit." It pits the four seasons of the calendar year and opposing players against one another in a "lightning-fast" game for survival. The object is to overpower your opponents and achieve the highest score by capturing their high-point "storm fronts" (or tricks).

The number of hands needed to win a game depends on the number of players. Most games can be played in less than an hour.

The Weatherslam card deck consists of fifty-six cards depicting violent storm systems—such as hurricanes and typhoons—and balmy weather conditions for each season of the year. Players vie, through competitive bidding, for the privilege of declaring "Seasonal Trump" (i.e., Spring, Summer, Autumn, or Winter). The season named trump has power over the other season in a hand of play.

The game includes a rule booklet and a supply of score sheets. It is recommended for ages seven to adult and costs $9.95.

How the Weather Works
Michael Mogil
522 Baylor Avenue
Rockville, MD 20580
(301) 762–7669

Michael Mogil is a professional meteorologist who has independently produced two filmstrips.

CHAPTER 15: HOW TO LEARN MORE

One is on tornadoes and is entitled *Stormin' All Over*. It costs $39.50 and also is available as a set of 35mm slides for $99.50. The other film is about clouds and is entitled *Clouding the Issue*. It sells for $35 and also is available on 35mm slides for $75. In conjunction with the National Weather Association, Mogil also has designed and created a set of three 1-by-2-foot cloud charts that are excellent for classroom use and as reference tools outdoors. The chart set costs $9.00 and also is available as a set of 35mm slides for $75. Shipping and handling charges depend on size of order.

Educational Filmstrips

The Everyday Weather Project
State University of New York at Brockport
Brockport, NY 14420
(716) 395–2352

The Everyday Weather Project includes a series of videotapes and sound filmstrips. Videotape titles include *Hazardous Weather: Hurricanes*, *Hazardous Weather: Thunderstorms*, and *The Sense of Weather*. These productions study storm formation and life cycles. All three videocassettes are priced at $74.95. Filmstrip titles are *Sensing and Analyzing Weather*, *Weather Systems*, *Weather Forecasting*, *Weather Radar*, *Weather Satellites*, and the filmstrip version of *Hazardous Weather: Hurricanes*. Filmstrips are $34.95 each, with a special package price of $174.95 for all six. Add shipping and handling charges of $5.00 to all orders.

Weather Experiment Kits

Hubbard Scientific, Inc.
P. O. Box 760
Chippewa Falls, WI 54729–0760
(800) 323-8368

(All kits below include a teacher's guide.)

Evaporation Kit: Demonstrate evaporation with this simple experiment. A single balance-beam scale with wire supports and sponge weights illustrates the factors that affect evaporation. The price is $15.

Air Mass Generator Kit: Demonstrate the formation of stable and unstable air masses with this informative experiment that effectively illustrates the relationship between temperature and air movements. This kit includes a plastic cylinder and track, two thermometers, a dish with cover, all required tubing, a funnel, and a smoke generator. The price is $26.

Cloud Chamber Kit: Here's a kit that allows you to see and trace radioactive elements. You'll be amazed as it reveals trails of radioactive rays. This kit includes a plastic cloud chamber, a radioactive source (not harmful, of course), and a magnet. The price is $9.00.

Coriolis Effect Kit: The Coriolis effect is perhaps one of the most difficult weather factors to explain, yet it determines the major weather patterns throughout the planet. This experiment includes steel spheres that track a pattern on the kit's turntables, simulating the effect of the Earth's rotation on winds, ocean currents, and material objects. The kit includes a 14-inch-diameter diameter base with erasable tracing surface mounted on the turntable, two $5/8$-inch-diameter steel spheres and a removable launcher for the steel spheres. The price is $25.

Weather Watch Board: This is a 35-by-45-inch, vacuum-formed board that can be mounted on a wall. It includes a radio that receives National Oceanic and Atmospheric Administration broadcasts. It also includes a set of fourteen study prints to help in interpreting observable weather conditions. The price is $277.

Weather Terminology

We close *The Weather Sourcebook* with a reference glossary of weather terms. Some are common, while others are pretty obscure. Still, our hope is that it will help you with some of the technical terms we've used throughout the book.

Air mass: A large body of air that has relatively uniform weather conditions, such as temperature, humidity, and pressure.

Air pressure: The force exerted by the weight of the air pressing down on the Earth. It decreases with increasing altitude because there is less air above to press down.

Anemometer: An instrument that measures how fast the wind is blowing. A simple anemometer has several metal cups on spokes that are attached to a shaft. The shaft is linked to dials. When the wind spins the cups, its speed can be read from the dials.

Aneroid: Not using liquid. An aneroid barometer has a needle connected to the top of a small box that has had some of its air taken out. A change in air pressure moves the box lid up and down, and this moves the needle. The air pressure can be read from the needle's position on a scale.

Anticyclone: An area of high pressure. The winds of an anticyclone blow in a spiral outward from the center.

Atmosphere: The layer of gases, or air, that surrounds the Earth. The atmosphere is made up mostly of the gases nitrogen and oxygen.

Barometer: A device to measure the pressure of the air. The U.S. standard is to measure air pressure in terms of the height of a column of mercury that would exert the same pressure. Thus, the force applied by standard sea-level pressure is the equivalent of the pressure that would be applied by a column of mercury 29.92 inches high. Some barometers are equipped to record the changes as they occur on a chart that depicts both air pressure and time.

Carbon dioxide: A colorless, odorless gas given out by animals when they breathe. This gas is taken in by plants, which give out oxygen.

Chinook: A warm, dry wind that blows down out of the Rocky Mountain slopes in winter and early spring.

Chlorofluorocarbons (CFCs): Chemicals made up from the elements chlorine, fluorine, and carbon. CFCs can damage the ozone layer.

Climate: An established, distinct pattern of weather conditions that exists in a place over a period of many years.

Cloud: Moisture in the air that has condensed onto particles of dust or smoke.

Condense: To change from a gas or vapor to a liquid. An example is when steam condenses into water.

Condensation: Moisture that comes out of air that has cooled to its saturation point and no longer can hold all the water vapor it contains.

Cyclone: A large mass of low-pressure air, with

winds that blow counterclockwise in the Northern Hemisphere and clockwise in the Southern Hemisphere.

Depression: An area of low atmospheric pressure. It is another name for a cyclone.

Dew: Moisture condensation that collects on unprotected objects outdoors when the air has cooled below its saturation point.

Drizzle: Precipitation droplets from stratiform clouds that are much smaller than those from other forms. The droplets do not intermix and combine into rain drops and still are small when they reach Earth.

Evaporation: The change from a liquid or a solid to a vapor or gas.

Foehn: A warm, dry wind that blows down out of mountain slopes in winter and early spring.

Fog: Clouds that form at ground level.

Front: A long, narrow band of changing weather that marks the area where two different kinds of air masses meet. We name it for the prevailing air mass. For example, when a cold air mass pushes a warmer air mass out of the way, we call the boundary between them a *cold front*. If neither of the two air masses dominates and the boundary stays still, we call it a *stationary front*. An unusual condition in which an advancing cold front overruns a warm front is called an *occluded front*.

Frost: Frozen moisture that has formed from condensation.

Hail: Large pieces of ice that fall to the ground from very large and cold thunderstorms.

High: A short name for an area of high pressure. It is also called an anticyclone.

Horse latitudes: Latitudes where sailing ships often encounter extended periods of calm winds. In this region ship's captains would order horses thrown overboard to preserve food and lighten the ships' loads. The many horse carcasses seen floating on the still sea there earned the region its name.

Humidity: The amount of moisture in the air in the form of suspended (and often invisible) water vapor.

Hurricane: A violent storm that is made by a tropical cyclone.

Hydrogen: A gas that is the lightest chemical element. It can float balloons, but it burns very easily.

Hygrometer: An instrument that measures the humidity of the air.

Ice age: A time in the Earth's past when large areas of the world's land were covered by thick ice. There were several ice ages, but only the last one is known as the Ice Age. This Ice Age took place during the Pleistocene epoch, from 2.5 million years to about 10,000 years ago.

Ice caps: Thick sheets of ice covering large areas of land, especially in Antarctica and Greenland. Sheets like these covered much larger areas during ice ages.

Isobars: Lines connecting points of equal atmospheric pressure on a weather map.

Isotherm: Lines connecting points of equal air temperature on a weather map.

Low: A short name for an area of low pressure, or a cyclone.

Magnetic field: The area of force that surrounds a magnet. A magnetic field surrounds Earth, but it is weakest at the poles.

Mercury: A silver-colored metal that is a liquid at normal temperatures. It is often used in thermometers and barometers.

Monsoon: A wind that changes direction according to the season. Also, the rains that it brings to certain parts of the world in summer.

Observatory: A place or building where people study the stars and planets. Observatories of today use cameras linked with high-powered tracking telescopes.

Ozone: A kind of oxygen whose molecules have three oxygen atoms instead of the usual two. In our atmosphere, ozone shields Earth from the sun's dangerous rays. On Earth's surface, ozone is harmful to all life.

Precipitation: Any form of water in the form of droplets or ice that falls to the earth, whether as rain, snow, sleet, or hail.

Radar: A device that bounces radio waves off an object in order to measure the distance to it and to keep track of its movements.

Radiation: The flow of particles and rays, such as light and radio waves. It also means the energy released from an atom.

WEATHER TERMINOLOGY

Rain: Precipitation that reaches the Earth as large droplets. It may begin as smaller droplets high in the sky; but as the droplets fall, they intermingle and combine to become larger droplets. We classify rain by the amount that falls and by the results of the free-fall mixing action, describing rain as light, moderate, or heavy.

Rain forests: Forests growing in tropical areas that have heavy rainfall.

Rain gauge: An instrument that collects precipitation, enabling weather observers to measure the amount.

Sleet: Rain that has developed a frozen outer coating but remains liquid in the center.

Snow: Snow is caused by the condensation of water vapor on particles of dust in the clouds at temperatures below the freezing point of water. Each snowflake has a unique shape. When viewed under a microscope, they appear as an endless variety of six-sided crystal structures.

Snow line: An imaginary line drawn to show the height above which the land is covered with snow year-round. The snow line is affected by a location's elevation and distance from the equator.

Sonde: An instrument for measuring and sending back information on weather or other conditions high above the Earth's surface. Sondes are sent aloft on weather balloons.

Static electricity: Electricity that does not flow. Instead, it builds up until it discharges in a spark. Current electricity is the type that flows, the kind used to light buildings.

Temperature: A scientific scale we use to determine the amount of heat energy contained in the air. The faster air molecules move, the higher the temperature measurement.

Tides: The regular rise and fall of the ocean levels, caused by the moon's pull on the water.

Tornado: A small but most violent storm averaging only about 300 yards across. Tornadoes are characterized by a dark funnel cloud with twisting winds spinning at up to 300 miles per hour.

Tree line: An imaginary line above which it is too cold for trees to grow.

Tropics: The zones around the Earth on both sides of the equator. The climate in these areas is very warm or hot all year round. Tropics often have heavy rainfall.

Vapor: Particles of moisture or solids that form clouds or smoke. Air can hold moisture that you cannot see.

Waterspout: A circular column of air caused by the rising of an overheated layer of air over a lake or an ocean that draws water or seawater up at the base of the column. Occurring mostly in tropical regions, waterspouts develop under unstable weather conditions and may damage ships.

Weather vane: Also called wind vane or weathercock. A device that pivots on a vertical shaft and points into the direction of the prevailing wind. Usually shaped like an arrow, the feathered end is pushed back by the wind so the head points into it.

Whirlwind: A circular column of air caused by the rising of an overheated layer of air near the ground. Occurring most frequently in the deserts, whirlwinds may carry dust and sand more than 1,000 feet above the earth.

Wind: Any movement of masses of air, described by stating the direction from which they blow. For example, a west wind moves *from* the west *to* the east.

Index

AAWO. *See* American Association of Weather Observers
Acid rain, 99
Adiabatic lapse rate, 82
Advection dew, 101
Advection fog, 101
Aerosol particles, 11
Air pressure, 2, 5, 211, 212
   Arctic highs, 129, 130
   and barometers, 55, 59–65, 131, 151
   effects of changes in, 55–56
   and "The Storm of the Century," 56–59
   of tornadoes, 47
   and wind direction, 72, 123–24
Airspeed, 137–38
Altitude, and temperature changes, 82
Altocumulus clouds, 112, 113
Altostratus clouds, 113
Amateur weather observers, 115–16, 147–48, 149. *See also* Forecasting; Home weather stations; Online services
American Association of Weather Observers (AAWO), 149, 151–52, 154, 169
American Meteorological Society, 148, 162, 166
American Weather Observer Supplemental Observation Network, 152
Anemometers, 78, 211
Animal behavior, and weather lore, 17–23
Anticyclones, 211. *See also* High-pressure areas
Arctic high, 129, 130
Arctic sea smoke, 101
Army Corps of Engineers, 167
ASOS. *See* Automated Surface Observing System
Atmosphere, 1–2, 211. *See also* Air pressure
Automated Surface Observing System (ASOS), 168–69
Aviation weather, 135–36
   aviation magazines, 143
   and the jet stream, 140–41
   and microbursts, 136–39
   online services for, 141–42, 144
   software for flight planning, 144–46
   videos, books, and courses on, 143

Backing wind, 72
Barkhans, 76
Barometers, 55, 211
   aneroid, 60, 63, 151
   barographs, 63–64, 131
   digital, 62
   fishing, 131
   invention of, 59–60
   marine, historical reproduction, 131
   mercurial, 60, 63, 151
   for travel, 64–65
   use, care, and calibration of, 60–61
Beaufort scale, 124
Blizzards, 48–51, 52, 56
Books about weather, 60, 79, 118, 143, 154
   for adults and teenagers, 187–92
   for children, 192–95
   coffee-table books, 195

Carbon dioxide, 1, 2, 10, 11, 12–17, 211
Cats, and weather lore, 18–21
CD-ROM, 205–6
Charts
   of cloud types, 112, 118
   hurricane tracking, 53–54
   slide chart, 183
   weather and oceanographic, 128
Chesapeake-Potomac Hurricane (1933), 32, 34–35
Chinook winds, 68, 211
Chlorofluorocarbons, 211
Cirrocumulus clouds, 111, 112
Cirrostratus clouds, 111, 112
Cirrus clouds, 111, 112, 114
Climate, 2, 211
   carbon dioxide and future of, 12–16
   and the greenhouse effect, 9–12
   plate-tectonics theory and, 8–9
   types of, 2–7, 213
   volcanic eruptions and, 8
   world data on, 14–15
Clouds, 97, 101, 137, 211
   classification of, 111–13
   color of, 110–11

WEATHER TERMINOLOGY

formation of, 110
and the greenhouse effect, 10–11
high and low pressure areas and, 55–56, 59
products relating to, 118–19
seeding of, 116–17
and short-term forecasts, 109–10
warm and cold, 99
Coastal Research Amphibious Buggy (crab), 167–68
Cold air, condensation of, 2, 55
Cold humid and polar climates, 4–5
Colleges and universities, 199–201
Columbus, Christopher, 126–27
Computer games, 14
Computers. *See* Online services; Software
Condensation, 97, 110, 211
Convectional rain, 99–100
Cotton region shelter, 153
Crickets, and temperature, 86
Cumulonimbus clouds, 113, 136
Cumulus clouds, 111, 113, 136
Cyclones, 211–2. *See also* Low-pressure areas; specific storm types

Damage estimates, U.S. weather disasters, 39–40
Deltas, 103
Depressions, 100, 212. *See also* Low-pressure areas
Deserts, 3, 6–7, 67, 76
Dew, 97, 100–101, 110, 212
Dew point, 84–85, 97–98, 100–101, 110, 212
Divine wind, 69–71
Doldrums, 67
Downdrafts and downbursts, 137–38
Drizzle, 98, 99, 102, 212
Dry climates, 6–7
Dry seasons, 5
Dune types, 76
Dust, 99, 111

El Chichón, 8–9
Elevation, and barometric pressure, 60–61
El Niño, 88–90
End moraine, 103
Equatorial climates, 5
Erosion, 76, 102–3, 167–68
Evaporation, 97, 136, 212
Experiment kits, 118, 209

FAA. *See* Federal Aviation Administration
Fax services, 54, 133
Feathered clouds, 111
Federal Aviation Administration (FAA), 61, 134, 141
Federal Emergency Management Agency (FEMA), 42
FEMA. *See* Federal Emergency Management Agency
Films and film strips, 207, 208. *See also* Videos
Foehns, 68, 212
Fog, 97, 100, 110, 117, 212
Folklore about weather
  and animal behavior, 22–23
  and cats, 18–21
  and clouds, 113–14
  the "divine wind," 67–71
  and forecasting by human senses, 24–25
  forecasting goats, 17–18
  Groundhog Day, 21–22
  and insect behavior, 23
  and plants, 23–24
  and "rainmakers," 116–17
  sailors' sayings about the wind, 72
  sayings from world cultures, 25
  and weather patterns, 25
Forecasting
  and animal behavior, 17–23
  by cloud formations and wind direction, 113–14
  and folklore about weather patterns, 25
  and human senses, 24–25
  and insect behavior, 23, 86
  major equipment for, 156–58, 171
  marine weather, general rules for, 121–23
  and plants, 23–24
  slide chart for, 183
  software for, 156
  by wind direction and air pressure, 72, 123–24
Fossil fuel consumption, 10
Freezing fog, 101
Freezing rain and drizzle, 99
Front, defined, 212
Frontal rain, 99, 100
Frost, 89, 93, 102, 212
Funnel clouds. *See* Tornadoes

Games, 14, 208–9
Glaciers, 103

INDEX

Glass sculpture, of cloudless Earth, 119
Global warming, 10–12, 13
    educational materials on, 14, 202
Goats, and weather lore, 17–18
Gravity, 55
Greenhouse effect, 2–3, 10–12
Groundhog Day, 21

Hail, 43, 97, 99, 101–2, 212
High cloud types, 111
High-pressure areas, 2, 55–56, 129, 130, 212
Hillenger Effect, 24
Home weather stations, 145, 150–51, 153–54, 157
Horse latitudes, 68, 125, 212
Humidity, 55, 61, 97, 212
    measurement of, 95–96, 98, 106, 107, 108
Humiture factor, 86–88
Hurricanes, 31–32, 71, 212
    Chesapeake-Potomac Hurricane, 32, 34–35
    classification of, 33
    and Columbus's voyages, 126–27
    computerized tracking program, 178–80
    El Niño and, 88–90
    Hurricane Andrew, 203–4
    Hurricane Bob, 161–62
    Hurricane Hugo, 162, 165, 204
    tracking chart for, 53–54
    videos about, 208–9
Hydrogen, 212
Hygrometers, 98, 106, 212
Hygrothermographs, 95–96, 107

Ice, 97, 99, 113, 114
Ice ages, 12, 103, 212
Icebergs, 103, 130
Ice pellets, 99
Infragravity waves, 167–68
Insects, and short-term forecasts, 23, 86
Instrument shelters, 78, 99, 153, 154
Isobars and isotherms, 212

Jet stream, 58, 59, 89, 140–41

Kamikaze, 70, 71
Kublai Khan, 69–70

Lightning sensing devices, 53, 118–19
Low cloud types, 111, 113

Low-pressure areas, 2, 5, 56, 112
    blizzards and, 48–51
    and the jet stream, 141
    "The Storm of the Century," 56–59
    typhoons and, 71
    See also Hurricanes; Tornadoes

Magazines, 132, 143, 206
Manual of Barometry (U.S. Government Printing Office), 60
Maps, topographical, 60, 154
Marine weather
    and the Beaufort scale, 124
    charts and, 128
    and Columbus's voyages, 126–27
    electronic information for, 133–34
    forecasting, general rules for, 121–23
    horse latitudes, 67, 125, 212
    marine barometers, 131
    sailing and yachting magazines, 132
    and surf zone research, 167–68
    and the Titanic, 129–30
    wind direction and air pressure and, 123–24
Meanders, 103
Meteorologists, 135, 162, 164, 165
    professional organizations for, 148, 166
Microbursts, 136–37, 143
Middle cloud types, 111, 113
Mississippi River, 103
Mist, 100
Mobile weather stations, 184
Moisture, 31, 97–98, 102–3. See also Clouds; Humidity; specific types of precipitation and storms
Monitoring and recording devices
    barographs, 63–64, 131
    for precipitation and humidity, 106–7, 108
    for temperature and humidity, 94, 95–96
    for wind, 78–80
    See also Weather stations
Monsoon climates, 5
Monsoons, 90, 212
Mountain climates, 7–8
Mountains, 82, 100, 102–3
Mount St. Helens volcano, 8, 9
Movie weather quiz, 26–29
Museum of American Weather (Haverhill, N.H.), 198–99

Mushroom rocks, 76

NASA. *See* National Aviation and Space Administration
National Aviation and Space Administration (NASA), 207–8
National Bureau of Standards, 85
National Council for Industrial Meteorologists, 166
National Hurricane Center, 165
National Marine Fisheries Service, 134
National Oceanic and Atmospheric Administration (NOAA), 42, 89, 134, 147–48, 162, 206
    radio weather broadcasts, 173–75
    See also National Weather Service
National Severe Storms Laboratory, 54
National Weather Association, 148
National Weather Service (NWS), 54, 134, 201
    Automated Surface Observing System, 168–69
    Observing Handbook No. 2, 154
    SkyWarn system, 115+16
    standards for barometers, 61
    standards for home weather stations, 150, 151, 153
    standards for thermometers, 84, 85
    and tornado warnings, 40, 160
    and The Weather Channel, 162, 165
    WS Form B-91, 154
Native American rain dance, 116
Navigation aids, marine, 133–34
Newsletters, 54–55
NEXRAD Doppler radar systems, 159–61
Nimbostratus clouds, 113
Nitrogen, 1
Nitrogen oxides, 13
NOAA. *See* National Oceanic and Atmospheric Administration
North Atlantic Drift, 82
NWS. *See* National Weather Service

Observer's Handbook (British Meteorological Office), 60
Ocean currents, 82, 88–89, 101
Oceanographic tables, 134
Online services, 134, 177–78, 180–85
    for aviation weather, 141–42
    computer bulletin boards, 173, 177–78
Organizations, 148, 149, 154, 158, 162, 166

American Association of Weather Observers, 148, 151–52, 154, 169
Oxygen, 1, 2
Ozone, 212

Pager service, 144–45
Plants, 1, 2, 23–24, 76
Plate-tectonics theory, 9–10
Precipitation, 98–99, 103–4, 150, 212. *See also* specific types of precipitation
Probes, temperature and humidity, 85, 98

Radar, 159–60, 212
Radiation fog, 101
Radio weather broadcasts, 174–75
Rain, 98, 99, 100–101, 150, 160, 213
    and climate types, 3–8
    and cloud seeding, 116–17
    size and shape of raindrops, 102
    world records for, 103–4
Rainbows, 113, 114–15
Rain gauges, 89, 107, 150, 153, 154, 213
Relative humidity, 98
Relief rain, 92, 100
Rivers, 102–3
Rotation, Earth's, 2

Sahara Desert, 76
Sahel, desertification of, 3
Sandstorms, 76
Saturation point, 97–98
Sea-level standard day, 142
Seif dunes, 76
Semiarid climate, 6
Seminars, 172
Sensing devices, for lightning, 53, 118–19
SkyWarn system, 115–16
Sleet, 97, 99, 103, 213
Slide chart, 174, 198
Sling psychrometers, 77, 105
Snow, 11, 89, 97, 99, 150, 160, 213
    world records for, 103, 104
Snow line, 213
Software, 14, 118, 133, 178–80, 181–82
    for aviation weather and flight-planning, 144
    for forecasting, 156
Sonde, 213
Specific humidity, 98

INDEX

Static electricity, 213
Storm chasers, 41, 73–75, 202–3, 204–5
"Storm of the Century" (March 1993), 56–59
Storm surges, 34–36, 57, 70, 167–68
Stratocumulus clouds, 112, 113
Stratus clouds, 111, 113
Sublimation, 97
Subtropics, 67
Sulfate emissions, 10–11
Sulfur dioxide, 13
Sulfuric-acid droplets, 9
Sunlight, 1, 2, 110–11

Telephone recorded forecasts, 176–77
Television, cable, 161–65
Temperature, 31, 60, 150, 154, 213
  and air density, 55
  and altitude, 82
  and carbon dioxide levels, 15, 16
  El Niño and, 88–90
  and geographic position, 82
  and heating of land and water, 82–83
  highest and lowest recorded, 91–91
  humiture and windchill factors, 86–88
  measurement of, 83–86
  and ocean currents, 82
  and solar radiation, 15, 81
  volcanic eruptions and, 8–9
Terminology, 211–13
Thermographs, 95–96
Thermometers
  and dew point determination, 84–85
  digital, 94, 95, 96
  historical reproduction, 94
  maximum-minimum, 83–84, 153
  placement of, 150, 153
  tests for, 84
Thunderheads, 113
Thunderstorms, 43–45, 46, 136, 160, 203
Tidal bore, 36
Titanic, 129–30
Topographical maps, 56, 154
Tornadoes, 42, 60, 136–37, 203, 205, 213
  Andover tornado, 73–75, 165–66
  characteristics of, 43–45
  safety during, 46–47
  warning time for, 159, 160
Toy tornado, 54

Trade winds, 5, 67
Tree line, 213
Tropical climates, 5
Tropopause, 140
Typhoons, 70, 71

Ultraviolet-light rays, 2, 117
U.S. Department of Energy, 10
U.S. Geographical Survey, 154
U.S. Geological Survey, 154
USA Today, 176

Valleys, 7, 101
Vapor, 2, 97, 110, 213
Veering wind, 72
Vertical cloud types, 111, 113
Videos, 143, 157, 202–5, 207–9
Volcanoes, 1, 8

Warm air, expansion of, 2, 55
Watches, 65, 78, 95
Water erosion, 102–3
Waterspouts, 43, 45, 213. *See also* Tornadoes
Weather, defined, 2
Weather Channel, 161–65
Weather-satellite imagery, 118, 133, 184, 203–4
Weather stations, 145, 150–51, 153–54, 157, 184
Weather vanes, 78, 79, 213
Whirlwind, 76–77
Whiteouts, 49
Wind, 31, 47, 101, 130, 213
  and air pressure prediction, 123–24
  anemometers, 78
  Beaufort scale and, 124–25
  Chinook winds, 68, 211
  the "divine wind," 69–71
  erosion by, 76
  monitoring and recording devices, 79–80
  and sailors' folklore, 72
  and weather patterns, 67–68, 110
  weather vanes, 78, 79, 213
  world records for, 71–72
Windchill factor, 86–88
Wind chimes, 78–79
Wind shear, 138, 139

Yardangs, 76